Women and Cartography in the Progressive Era

In the twenty-first century we speak of a geospatial revolution, but over one hundred years ago another mapping revolution was in motion. Women's lives were in motion: they were playing a greater role in public on a variety of fronts. As women became more mobile (physically, socially, politically), they used and created geographic knowledge and maps. The maps created by American women were in motion too: created, shared, distributed as they worked to transform their landscapes.

Long overlooked, this women's work represents maps and mapping that today we would term community or participatory mapping, critical cartography and public geography. These historic examples of women-generated mapping represent the adoption of cartography and geography as part of women's work. While cartography and map use are not new, the adoption and application of this technology and form of communication in women's work and in multiple examples in the context of their social work, is unprecedented.

This study explores the implications of women's use of this technology in creating and presenting information and knowledge and wielding it to their own ends. This pioneering and original book will be essential reading for those working in Geography, Gender Studies, Women's Studies, Politics and History.

Christina E. Dando is Professor of Geography at the University of Nebraska Omaha. She received her B.A. in Geography and English from the University of North Dakota and her M.S. and Ph.D. in Geography from the University of Wisconsin-Madison. Her research interests include the impacts of media and technology on human perception and interaction with the environment, particularly the Great Plains. She is also interested in gender and geography, how landscape and environment have long been gendered as well as how gender impacts human experience and interaction with the environment. She is a member of the American Association of Geographers and of the Society of Woman Geographers. When not researching and writing, she enjoys exploring new landscapes and reading for sheer pleasure.

Studies in Historical Geography
Series Editor: Robert Mayhew

Historical geography has consistently been at the cutting edge of scholarship and research in human geography for the last fifty years. The first
generation of its practitioners, led by Clifford Darby, Carl Sauer and Vidal
de la Blache presented diligent archival studies of patterns of agriculture,
industry and the region through time and space. Drawing on this work, but
transcending it in terms of theoretical scope and substantive concerns, historical geography has long since developed into a highly interdisciplinary
field seeking to fuse the study of space and time. In doing so, it provides
new perspectives and insights into fundamental issues across both the humanities and social sciences. Having radically altered and expanded its conception of the theoretical underpinnings, data sources and styles of writing
through which it can practice its craft over the past twenty years, historical
geography is now a pluralistic, vibrant and interdisciplinary field of scholarship. In particular, two important trends can be discerned. First, there has
been a major 'cultural turn' in historical geography which has led to a concern with representation as driving historical-geographical consciousness,
leading scholars to a concern with text, interpretation and discourse rather
than the more materialist concerns of their predecessors. Secondly, there
has been a development of interdisciplinary scholarship, leading to fruitful
dialogues with historians of science, art historians and literary scholars in
particular which has revitalised the history of geographical thought as a
realm of inquiry in historical geography. Studies in Historical Geography
aims to provide a forum for the publication of scholarly work which en-
capsulates and furthers these developments. Aiming to attract an interdisciplinary and international authorship and audience, Studies in Historical
Geography will publish theoretical, historiographical and substantive contributions meshing time, space and society.

www.routledge.com/Studies-in-Historical-Geography/book-series/
ASHSER-1344.

Women and Cartography in the Progressive Era

Christina E. Dando

Routledge
Taylor & Francis Group

LONDON AND NEW YORK

First published 2018
by Routledge

2 Park Square, Milton Park, Abingdon, Oxfordshire OX14 4RN
52 Vanderbilt Avenue, New York, NY 10017

Routledge is an imprint of the Taylor & Francis Group, an informa business

First issued in paperback 2019

British Library Cataloguing-in-Publication Data
A catalogue record for this book is available from the British Library

Library of Congress Cataloging-in-Publication Data
A catalog record for this book has been requested.

ISBN: 978-1-4724-5118-7 (hbk)
ISBN: 978-0-367-24530-6 (pbk)

Typeset in Times New Roman
by codeMantra

Dedicated to
my grandmothers Myrtle Foster Dando and Bernice Bartowick
Zaporowsky
my mother Caroline Zaporowsky Dando
and my daughter Emmaline Sophia Dando Sabin

Contents

List of figures

List of tables

Acknowledgements

This project has been midwived by many. It had its origins at the History of Cartography Project at the University of Wisconsin-Madison. Working on the project and engaging in conversations with David and Roz Woodward got the wheels turning. While I wasn't yet in a position to pursue this work, the nurturing environment of the project as well as the map-obsessed atmosphere supported me and cultivated my interest in women and cartography.

It had its baby steps at a National Endowment for the Humanities Institute on Popular Cartography and Society at the Newberry Library, Chicago in July 2001. Thanks to Jim Akerman, I began to consider women and mapping, beginning with women and bicycling.

It continued with a Best Fellowship at the American Geographical Society Library, University of Wisconsin-Milwaukee in 2005, allowing me to begin work on women and the bicycle and where I encountered my first suffrage map.

As the project progressed, portions of chapters were developed through presentations at the American Association of Geographers Annual Meetings, beginning in 2002. In addition, versions were presented at the 2003 International Conference on the History of Cartography in Boston MA and Portland ME; the "Geography and Its Publics" Symposium at the 24th International Congress of History of Science, Technology, and Medicine in Manchester UK (2013); and the 2015 International Conference of Historical Geographers in London.

Portions of chapters 2, 4, and 5 were previously published as:

Dando, C. 2007. Riding the wheel: selling American women mobility and geographic knowledge. Media spaces, mediates places issue of *Acme: An International E-Journal for Critical Geographies* 6, 2: 174–210.

Dando, C. 2010. The map proves it!': map use by the American woman suffrage movement *Cartographica* 45, 4 (December): 221–240.

Dando, C. 2014. Hull house geography. *Geographers Biobibliographical Studies* 33: 137–165.

My community in Omaha has been very supportive. My department, the Geography/Geology Department of the University of Nebraska Omaha (UNO), has supported my work, especially my very understanding chair Bob Shuster. I have been lucky to have good friends in a writing group at UNO, the Women's Book Collective, especially Barbara Robins. The staff of the Criss Library, especially Mark Walters, the Interlibrary Loan Coordinator, were a tremendous help. And I am grateful for the neutral writing space of the coffeehouse Caffeine Dreams and Becky who keeps me caffeinated.

I am also very grateful for the camaraderie of my map history community friends – Matthew Edney, Jude Leimer, Penny Richards, and Judy Tyner. And my "September Merzavit's" – Marie, Pamela, Chris, and Brian – your support and encouragement mean a great deal to me.

My parents, William and Caroline Dando, have provided emotional and, at times, editorial support. They raised me with the conviction that women were equal to men, that women's history was just as important as masculine history, and encouraged me to pursue my dreams, regardless of societal gender norms and expectations. Not long ago, I was asked when did I become a "feminist geographer"? Thanks to my parents, I have always been a feminist geographer. For this, I am ever grateful.

Finally, a special thanks to my daughter Emmaline who assisted with the research and encouraged my work. It is exciting to see you become politically active – "woman's work" will continue and in good hands!

1   The power and the pleasure

In 1931, the Standard Oil Company distributed a road map with its cover featuring a woman behind the wheel of a car with a map unfolded before her, smiling as she looks at the viewer, as if pausing in planning her route (Figure 1.1). Images like this, of women reading maps, led me to ask: "when *did* women become part of the map reading audience?" Through my work in geography and the history of cartography, I had never seen nor heard of women as part of the audience for cartography. Cartography was the "science of princes," employed by empires, governments, and commercial interests (Harley 1988b, 281). The popular image of cartography portrayed in road map art, such as this image, depicts both men and women using maps with pleasure (Dando 2002). I had to wonder when this mainstream use of cartography began and when women became map-users. But in exploring the possibility of women as map-users, I inadvertently found women map-*producers* in the Progressive Era. The Progressive Era is generally considered to be from approximately 1890 until 1920: a time where both men and women were concerned with social and political reform and where women played a significant role. Prior to 1890, women had slowly been moving towards social reform efforts while associated with home and the private/domestic sphere, in particular, working on issues such as women's suffrage and temperance. In the 1890s, women took on a greater public role, still associated with home and domestic issues, working within women's organizations such as women's clubs to bring about social change (Rynbrandt 1997, 203–4). The "new woman" had education, possibly worked outside the home, had fewer children, and became involved with a wide range of activities, from charitable to political (McCammon et al. 2001, 53). By focusing on issues considered to be "municipal housekeeping," such as education, sanitation, and food safety, women could work towards improving their cities and society while still being perceived as "proper" women and mothers. Underlying their work was the belief "all municipal problems had to be solved before the city would be a good place in which to live" (Flanagan 1990, 1045). Further, the women's clubs and organizations advocated for the women to investigate for themselves the problems in their own communities, whether it was garbage problems or dance halls (Flanagan 1990, 1047–8).

Figure 1.1 "See California and the Entire Pacific West," Standard Oil Company of California, 1931. Courtesy of The Newberry Library, Chicago, IL (RMcN AE 007.14).

But, in order to address the issues they identified, women had to address the political structures governing their communities. Through women's clubs and organizations, women gained experience in political activism well before they attained the right to vote. Geographic knowledge and maps were an important element of their activism.

In this introduction, I will begin by considering the presence of geography in American culture and in education leading up to the Progressive Era. I will then consider the traditional practices of geography and cartography and their association with power and control. Their formal practice was largely associated with government and commercial interests. As a result, much of the scholarship in the history of geography and cartography focuses on these practices. But geography and cartography were used for more mundane purposes and everyday Americans, male and female, had a practical understanding of them. Recent scholarship on public geography, cartographic culture, critical cartography, and counter-mapping provides a framework for considering the practices of geography outside of the halls of power. Women have been involved in cartography for centuries, but their contributions were not acknowledged until work by women scholars over the last forty years brought them to light. As evidence has amassed, we now acknowledge their practice of geography was not unusual, but rather represents a previously overlooked group of practitioners. In approaching this practice of women's geography and cartography, I draw on the work of Donna Haraway and Bruno Latour to unpack the networks that were operating and how they employed their studies and maps. I conclude with a brief overview of the chapters, each centered on a different practice of geography and cartography by women in Progressive Era America.

Geography in American culture and women's education at the start of the Progressive Era

Geography and maps were widely available to the American public in the nineteenth century, with education including a healthy dose of geography for boys and girls. Jedediah Morse's *American Geography* (first published in 1784, but reprinted over 50 times) was a common fixture in American homes (Jerabek 1943; Withers 2006, 724–5).[1] Beyond the schools, geography texts and maps were widely available for all ages, even in the form of games and puzzles (Brückner 1999, 319).[2] Magazines, both for general readership and for a woman audience, included geographical articles and maps among their pages (Dando 2003). Newspaper articles included maps, ranging from local to international. Both men and women read books on history, biography, travel literature, and fiction, while there were some gender-specific exceptions, such as men reading agricultural manuals and women recipe books (Kelley 1996, 404). As individuals traveled, by train, by bicycle, they used guidebooks and maps to find their way. Geography and maps were part of American culture and Americans had a "cartographic culture," an "understanding of and attitudes towards maps as representations of spatial knowledge" (Edney 1994, 385).

This cartographic culture began with education. Women's education in the United States expanded dramatically after the American Revolution, with hundreds of female academies and seminaries founded (Kelley 1996, 404). Women's academies and seminaries such as Willard's Troy Seminary (1821), Mount Holyoke Seminary (1836), and Rockford Female Seminary (1847) educated young women and provided them with classical education, emphasizing subjects such as Latin, Greek, mathematics, rhetoric, history, modern languages, physical sciences, and geography. Mary Kelley's work on women and reading documents several instances where women discussed their geography studies in diaries and letters, such as Harriet Haynes being presented with a book for achieving "the highest status in Third Class in Geography" (Kelley 1996, 409–410).[3] In the early

nineteenth century, the Mordecai family is documented by Penny Richards as engaging in education on a variety of levels, from educating their
own children, to tutoring others' children, to running a school, and "At
the Mordecai school, geography was one of the first subjects to which a girl
was introduced" (Richards 2004, 4–5). Older girls took "advanced" classes
in geography.

Geography was considered, from the late eighteenth century in the United
States, an appropriate education topic for girls and women: "Both geography and map literacy were considered subjects that cultivated literacy as well
as preparing women for 'the trivial conversations of social circles'"(Schulten
2012, 18). Other scholars believe geography was taught because it "could
instill habits of good citizenship, develop national pride, and create public
support for surveys and scientific expeditions" (Tolley 2003, 14).

For whatever the reason, geography and cartography were common elements of girls/womens education, both as a stand-alone topic and as integral
to other subjects. Emma Willard, who sold over a million textbooks between
the 1820s and 1860s, integrated geography and maps into her approach in
history (Schulten 2012, 22). In addition to writing textbooks and creating
innovation teaching methods, such as the 'Temple of Time,' Willard was a
teacher and founder of the Troy Female Seminary, "one of the country's fin-
est institutions of female education"(Schulten 2012, 18–20). Troy's students
drew maps and wrote about geography as they practiced grammar and pen-
manship and learned geography, history, and astronomy. Willard's curriculum and that of Mount Holyoke were often used as a template for other
female academies. In the early nineteenth century, seminaries balanced academic subjects with household subjects such as needlecraft: Penny Richards
suggests "women may have become familiar with precise, graphic, spatial
information through their routine experiences using, altering, and copying
dress patterns; quilt planning and needlework may also have involved the
use of such information" (Richards 2004, 12).

Through the nineteenth century, women's educational opportunities
grew, with increasing numbers of women having access to higher education. Some of these opportunities were self-created, with women founding
women's societies, classes, and clubs, where they read broadly, wrote com-
mentaries and poems, and held debates (Kelley 1996, 421). Jedidiah Morse
claimed his *Elements of Geography* (1795) was useful not only to learn geography but also as a reading book in schools and for leisurely reading at home
(Tolley 2003, 18). Eventually, women's colleges were established: Vassar in
1865, Smith in 1875, Bryn Mawr in 1884, and Mount Holyoke in 1888. Agnes
Holbrook, one of the architects of the Hull-House maps, was educated at
Wellesley College (founded in 1870) where her coursework included statistics and art, well-preparing her for translating data into graphics, such
as maps (Dillon undated). Outside of college coursework, "science clubs"
brought students together to read and debate the latest in scientific advances
(Brown 2004, 85; Stebner 1997, 121).

While young women had access to higher education, they did not have many opportunities to apply their knowledge, other than to be wives, mothers, and teachers and perhaps, for a select few, Christian missionaries (Townsend 1990, 84). In 1889, Ellen Gates Starr, one of the founders of Hull-House, wrote to a friend:

> I pity girls who have nothing in the world to do. People get up in church and in missionary meetings and tell them about the suffering in the world and the need of relieving it, all of which they knew and it ends in them giving some money which isn't theirs as they never earned it and all their emotion over it and their restlessness to do something has to end in that... . [N]o body ever shows them a place and says 'There do this.' I know that girls want to do. I have talked with enough of them poor little things!'
>
> Starr quoted in Jackson 2000, 46

At the time she wrote these lines, Starr was founding Hull-House with her good friend from college, Jane Addams, creating opportunities for themselves to apply their educations to meaningful work. With the Progressive Era, educated women began to apply their knowledge to their social reform work.

Hegemonic geography and cartography

> Geography begins only when geographers begin writing it.
> Wooldridge and East 1951, 161

Academic geography became established at American institutions of higher education during the Progressive Era. Despite the presence of geography in American culture and its presence in women's educational institutions, it was just becoming a recognized academic subject in the United States.[4] William Morris Davis was appointed instructor of physical geography at Harvard University in 1878. By the turn of the twentieth century, there were only three professors of geography at American universities. The first department of geography in the United States was founded at the University of Chicago in 1903 (James and Martin 1981, 280–1 and 294). Despite its relatively late establishment in academia, geography has long been an essential field of knowledge, tied particularly to imperialism.

Geography as a realm of knowledge has existed at least since the ancient Greeks and based on an understanding of the world as the home of humans. Over time, it developed into an academic discipline focused on spatial information, involving methods that collect and process this information, such as fieldwork and mapping. Maps are viewed as *the* essential tool of geographers, being the primary but not the only method of analyzing as well as conveying spatial information (Cosgrove 2008, 1; Heffernan 2009, 11; Perkins 2009b, 389).

In considering the relationship between geography and maps, we have to acknowledge both the practice and the artifact: mapping is an aspect of active geographical practice, while maps are a means of communicating geographical knowledge. As Matthew Edney writes:

> The production of knowledge depended not only on the measured survey and observation of the landscape but also on the reconstitution and interpretation of the resultant data to create a single corpus of geographic knowledge.... . For geography, rational thought would bring all collected knowledge into a single archival corpus: the map.
>
> Edney 1994, 386–7

Mapping is not the only expression of geographical description; textual forms of geography have long dominated the field. Early Greek geographers' expression of geography was as written descriptions of their world. Later geographers employed both written and visual forms of geography (Cosgrove 2008, 7). Withers argues maps are "the most prevalent form of geographical writing" (Withers 2011, 42).

Geography and cartography are tied to imperialism, power, and control, with maps communicating "an imperial message ... used as an aggressive complement to the rhetoric of speeches, newspapers and written texts" (Harley 1988a, 57). This imperial geography involves the control of peoples, resources, territory, and knowledge, from a position of absolute power. Not surprisingly, much of geography and cartography's history is associated with masculinity (Huffman 1997; Rose 1993). Maps as the tools of empire are complicated artifacts, embodying art as well as technology, communicating as much as concealing, and imbued with power. For the most part, cartography has focused on technological advancements, on improved accuracy, of innovations in production, of changing formats of delivery, with the assumption of a world that "can be objectively known and faithfully mapped using scientific techniques that capture and display spatial information" (Kitchin et al. 2013, 480–1; Perkins 2009a, 126). This focus on advancements results in the creation of a narrative that develops smoothly from ancient Greece through the European Renaissance, and eventually to North America (Edney 2011, 332; Withers 2011, 42). In focusing on "progress," "the many simple, derivative maps were dismissed from consideration as inadequate or abnormal" (Edney 2011, 331).

In his benchmark essay, "Cartography without 'progress,'" Edney considers the practice of cartography and its portrayal in scholarship on the history of cartography. Just as cartography is perceived to have progressed smoothly through history, so too it advanced "smoothly" from an 'art,' with decorations filling in blank spaces on the maps, to a 'science,' with blank spaces kept as signifiers of "factual neutrality" (Edney 1993, 56). Edney argues, rather than an unbroken flow of progress, we need to acknowledge the distinctions between different practices of mapping, what he terms

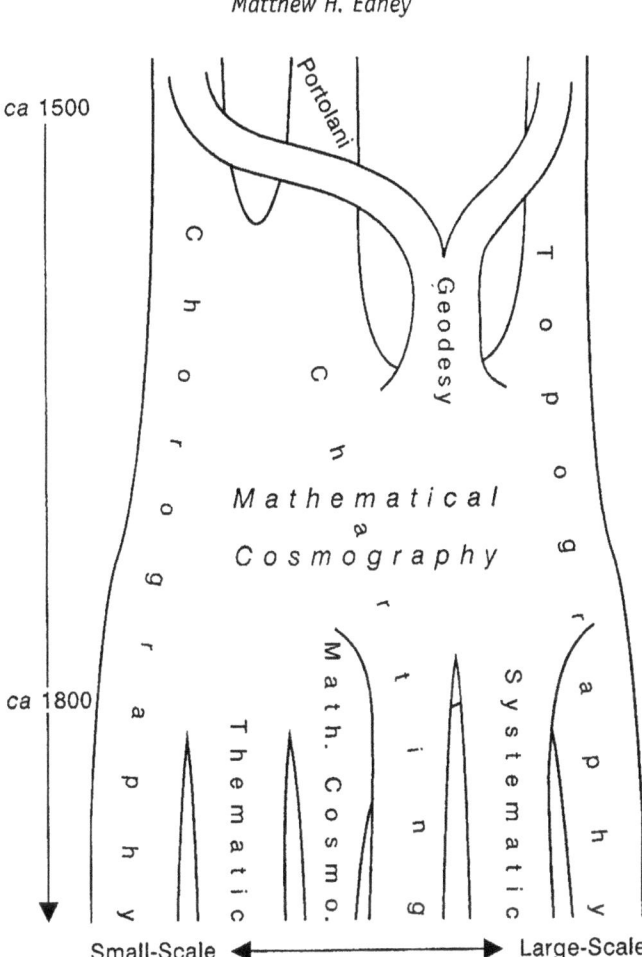

Figure 1.2 Schematic diagram from Edney 1993. By permission of Matthew Edney.

"cartographic modes."[5] Cartographic modes are "sets of cultural, social, and technical relations which define cartographic practices and which determine the character of cartographic information" and each mode is enmeshed in cultural, social, and technological relations (Edney 1993, 54 and 57). Thus, the mode reflects both the cartographic form and function and "the external raison d'être of the map" (Edney 1993, 58). Modes include a variety of mapping practices, such as chorography, thematic, mathematic, charting, systematic, and topographic, which have evolved and re-evolved over time. In explaining the modes, Edney employs a "schematic diagram of the eighteenth-century convergence of the formal cartographic modes and their

subsequent divergence after about 1800" (see Figure 1.2).[6] The diagram captures the flows of the different cartographic modes over time, as they split,
merge, converge, and split further, a braided stream of cartographic practice.
But this stream only reflects surficial flow. When scholars compare and contrast different modes of mapping, for example thematic maps versus nautical
maps, the differences are not just a matter of form but also incorporate "the
requirements of different social organizations for graphic representations to
aid their understanding of the human world" (Edney 1993, 57). The resulting
history of cartography captures this braided stream, reflecting "the history
of the creation, internal change, interaction, merging, bifurcation, and over-
lapping of cartographic modes" (Edney 1993, 58). Ultimately, Edney suggests
cartographic products are wrapped in the cultural, social, and technological
relations of their mode, as is the practice itself (Edney 1993, 58). However,
Edney focuses his essay only on, as he himself describes it, "'formal' cartography, which is prosecuted within the commercial and governmental con-
fines of the modern capitalist state" (Edney 2011, 306).

Formal cartography is hegemonic cartography, linked with control of
space and knowledge, and with the rise of capitalism (Firth 2014, 158–9).
They have been:

> … critical tools for the modern state and its agencies in shaping social
> and moral spaces, and they played a central role in the Western physi-
> cal and intellectual colonization of territories, peoples and the natural
> world. For over two centuries thematic and statistical maps have ex-
> tended these roles in supporting the bureaucratic concerns of the mod-
> ern state.
>
> Cosgrove 2008, 155

Maps were critical not only for comprehending distant lands, but also for
their role in nurturing a "sense of imagined community that was enor-
mously powerful in fostering the modern development of the nation-state"
(Perkins 2009b, 394).

Maps were such "critical tools" that they were "state secrets." J. Brian Harley in his classic essay "Silences and Secrecy" documents how some states
and their rulers pointedly kept their maps secret, fearing their knowledge
might be used against them (Harley 1988a). Navigation maps were tightly
controlled, tied to maintaining their commercial monopolies. Portugal and
Spain, in order to preserve their exclusive hold on their colonies, strictly regulated both the nautical charts and the maps of the new lands. Further, Harley
points out "silences" on maps are just as important as presences; that there
truly is no such thing as "empty space" on a map. An example of this would
be how many early European maps depicted the Americas as blank or empty,
reflecting political and cultural privileging of European geographic names
and knowledge and the "silencing" or "erasing" of the presence of Native
Americans and their geographic names and knowledge (Harley 1988a, 66).

With the proliferation of statistical data in the nineteenth century came a proliferation of maps "adapted as tools to make sense of particular kinds of information" (Schulten 2012, 3). Maps were used in new ways: to foster national unity, to map and control disease, to map social phenomena such as poverty or slavery and advocate for change, to promote economic development at home and abroad. While there was new data and new ways of presenting it, geography and cartography were still the tools of those in power. Some of the first thematic maps created were statistical in nature, mapping out education, poverty and other "moral" issues in France in the 1840s (Friendly and Palsky 2007). But as they mapped out these moral issues, they framed the issues through the cartographic form. In 1826, Charles Dupin created one of the earliest choropleth maps illustrating primary education, where the shading of light to dark reflects the number of children being sent to school, with lighter associated with higher levels and darker with lower, thus "The shading gave the impression of a light cast on the map, comparable to the light of knowledge … the scale of shading always transcribed a scale of values, always with sense that darker meant worse" (Friendly and Palsky 2007, 241). Susan Schulten documents the powerful map created from 1860 census data of the population density of slavery in the American South and how it was both a "landmark cartographic achievement and an eminently practical military instrument," copied and reproduced widely, and interpreted in numerous ways (Schulten 2012, 119–55). In particular, President Abraham Lincoln was particularly struck by the map and what he believed it revealed as he dealt with the Civil War. Lincoln's interest was significant enough to be recognized by the painter Francis Bicknell Carpenter who included the map in his image of *First Reading of the Emancipation Proclamation of President Lincoln*, painted in 1864 (Schulten 2012, 119).[7] Using the combined power of numbers and graphics, those in power had new tools to direct the course of the nation.

In the United States, the American Geographical Society (AGS) began as the "American Geographical and Statistical Society" in 1854, with the objective of collecting and disseminating information (Morin 2011, 33). Early members of the AGS were entrepreneurs, in the foreign service, in publishing, in the earth sciences, in the clergy, and in law and politics and all interested in the advantages such knowledge afforded, largely commercial (Morin 2011, 31). Between the United States Government and the AGS, statistics were focused largely on data and maps addressing U.S. business interests at home and abroad, particularly economic conditions. Some demographic studies and mapping were conducted using U.S. census data (Morin 2011, 36). Between the focus on exploration and the focus on economic development, the work of the AGS followed a "colonial science model – or more specifically, a political-economic imperial model of non-territorial domination over global regions through enumeration and comparison of raw materials, ultimately to benefit domestic manufacturing and industrialization" (Morin 2011, 37). The society dropped "Statistical" from its name in 1871 and shifted their focus towards exploration (Morin 2011, 35).

This focus on exploration was echoed by the creation of the National Geographic Society (NGS) in 1888. Initially founded by prominent members of the U.S. government, academic geographers, and business leaders, such as John Wesley Powell, William Morris Davis, Grove Karl Gilbert, and Henry Gannett, NGS was intended to be a "serious" scientific geography organization (Lutz and Collins 1993, 20).[8] But internal divisions over the direction of NGS toward a more general readership led to Davis forming the Association of American Geographers in 1904 expressively for professional geographers (Schulten 2001, 49). Geography struggled to be perceived as "scientific" and "academic" and worked to distance itself from the practices associated with school geography, i.e., description and rote learning, similar to its experiences in other countries, such as England (Maddrell 2009, 330).

Over the course of the nineteenth and early twentieth centuries, maps and mapping of all modes and the creation of geographic knowledge increased dramatically, conducted by many stakeholders. While research and mapping by the U.S. government, by geographical societies such as the AGS and NGS, and by academic geographers, can all be described as hegemonic, other work was taking place. Thematic mapping came to be seen as a critical tool during the Progressive Era by community activists. Surveys and studies of communities and the resulting maps were used to create "an ecological portrait of community in place, a foundation for its rational planning, and a stimulus to the civic virtue and community participation of its people" (Cosgrove 2008, 166).

Public geography/cartographic culture

Geography and mapping are not confined to governments and academia: all peoples have their own practice of geography and cartography. Recent work in the history of geography and cartography has termed the practice of geography outside of academia "public geography" and an understanding and use of maps in a culture as "cartographic culture." But much of the discussions of public geography and cartographic culture occur in different locations; in the scholarship on the history of geography and in work on the history of cartography. In approaching women's work in the Progressive Era, I believe both of these concepts are relevant. Both geography and cartography have production and dissemination, both have their practice in daily life, both have their "cultures." Can we draw a line between these concepts?

Public geography is the practice of geography outside of the academy and "put into the 'service' of the public" (Morin 2011, 12; Withers 1999). Charles Withers has conducted extensive work on public geography in eighteenth and nineteenth century England (Withers 1999, 2011,). English citizens' imaginations were captured by accounts from explorers and fired by the images from magic lantern shows (Smith 2013, 190). In parlors and at public lectures, citizens discussed, debated, and engaged with matters of geography. Geography was "situated as polite public knowledge in the public sphere"

(Withers 1999, 67). In Withers' work, he considers "some of the ways in which, over time and in different ways, geography has been undertaken" and that there is not a single way of "doing" geography but rather a wide range of practices of geographical writing and reading (Withers 2011, 39 and 41).

In the United States, Karen Morin's "social geography" of Charles Daly, long-term President of the AGS, documents his extensive practice of public geography through his various activities (Morin 2011). Daly was never an academic geographer but was nationally influential through his popular annual lectures, his promotion of exploration and American business interests, and his Congressional testimony. Daly's public geography took place at a variety of locations, only one of which was tied to traditional locations of power (Congress).

Withers writes historically the public practice of geography had:

> … a geography: of *sites* of production, consumption and negation … of experimental sites and *spaces of display*; of *private spaces* of polite conversation about experimental knowledge; of *the circulation of knowledge* between practitioner, author and audience; and of *practitioners themselves* on their itinerary. Such matters offer the possibility for exploring not just the situated geography of science as a form of public culture and, thus, the geography of the public sphere, but also the place of geography itself as a form of polite and public knowledge.
>
> Withers 1999, 49–50[9]

Withers' acknowledgement of geography's geographies is echoed in Livingstone 2000. But, whether academic or public, "geography's practices of depiction are always also about authority, authoritativeness and (more assumed than tested) 'accuracy' and 'authenticity' in different contexts" (Withers 2011, 43).

Withers' work is often paired with Hayden Lorimer's who examines the practice of geography in "non-traditional arenas," such as the Glenmore Lodge in Scotland where residential field courses were held for teenagers (Lorimer 2003). Lorimer's work focuses on bringing "greater recognition of knowledge production among geography's 'grass-roots' practitioners," calling for "telling small stories," which present both opportunities and challenges (Lorimer 2003, 200). The phrase "telling small stories" has become a catch-phrase for acknowledging the contributions of those who operated outside the parameters of traditional or academic geography, often using more biographical methods of researching their stories, such as interviews, papers, obituaries, oral histories and other resources to piece together their stories (see, for example, Maddrell 2009, 16–19). Through the work of scholars such as Withers, Morin, and Lorimer, a much broader picture of the public practice of geography comes into focus.

Similarly, we might say, at the beginning of the Progressive Era in the United States, the population had developed a cartographic culture … or perhaps a number of cartographic cultures. The origins of the term "cartographic culture" are fuzzy, with the term beginning to circulate in the early 1990s in the work of writers who contributed to the first volume of the *History of Cartography* series and their students.[10] Matthew Edney provides the clearest definition of the term in *Mapping an Empire*, his examination of the British systematic mapping of India:

> 'Cartographic culture' encompasses not material map artifacts but the understanding of the practice of cartography which a society possesses, the forms of representation employed to experience and explore the world, and the means whereby the social order permeates those representations in order to recast and recreate itself.
>
> Edney 1997, 36

In another essay, Edney explores the cartographic culture of the later Enlightenment in the United States, using an unfinished manuscript by Benjamin Vaughan. Vaughan's manuscript was a response to a proposal for a prime meridian to run through Washington DC and explicitly discusses the nature of maps and prime meridians, reflecting "later Enlightenment elite cartographic culture" (Edney 1994, 386). Edney's work situates Vaughan's work in a culture where an understanding of maps, mapping and cartography were common knowledge for American intellectuals and tied to political discourse.

Approaching cartographic culture from a different angle, Penny Richards has documented the use of emotional cartographic culture in "'Could I but Mark Out My Own Map of Life': Educated Women Embracing Cartography in the Nineteenth-Century American South" (Richards 2004). Using letters, maps, school records, and other archival materials, Richards explores how educated Southern women "turned to the cartographic form with familiarity and an expectation of solace. They used geographic vocabularies and metaphors to express their sense of isolation and their need for connection" (Richards 2004, 2). According to Richards, Edney's definition of cartographic culture expands the history of map use and "brings lay map use, domestic map use, and necessarily, women's map use into clearer view" (Richards 2004, 2). She documents how women, such as the Mordecai sisters, studied geography as part of their education, used geographical and cartographical turns of phrase in their correspondence to each other, and followed family members' travels on maps and atlases (Richards 2004, 6). Similarly, Richards found Priscilla Brownrigg Bailey employed maps to follow her husband's travels and, later in life, received maps from her granddaughter and her son, her son's map a hand-drawn visit to her much missed hometown, replete with familiar names, and even "evocative tastes and smells" (Richards 2004, 10). Richards concludes with a call to consider "all the readers in the home," that work is needed "approaching the availability of maps to users, the training of map users, and their attitudes towards maps," with the "sticking point" being the availability of evidence (Richards 2004, 11).

So often, in academic circles or in publishing books, scholars focus on the big important maps or mapping projects, which confine our view of cartography to the "science of princes," to those maps created by explorers or governments. By focusing on "important maps," we forget "mapping is an everyday activity and tool," used throughout world, throughout people's lives, and is in many ways mundane, as they create and use maps to navigate their everyday lives (Perkins 2009b, 385).

Richards' work is one of the first to consider non-hegemonic map users. Studies examining maps produced "by and for peoples marginalized within Western societies remain rare" with a handful of studies produced on Native American map-making, and a little on Black American mapping and women's mapping (Hanna 2012, 53). Edney acknowledges despite evolving conceptualization of what constitutes a "map" and work to advance the ways maps function on social levels, "there was little interest in developing any sense of the larger picture of how map making and map use *have* functioned and developed" (Edney 2011, 333).[11] And there continues to be calls to examine the various "cultures" of map use, as in the work of Chris Perkins who calls for "a broad, contextual and social approach to map use" (Perkins 2008, 150) with attention shifting onto "processes, institutions, social groups, power, interactions between different elements in networks, emotions at play in mapping, the nature of mapping tasks and a concern with practice" (Perkins 2008, 152). Perkins, in considering "Cultures of Map Use" writes in 2009: "Mapping is moving to new sites, no longer under the control of cartographers" (Perkins 2008, 152).

Critical cartography is a relatively new concept used to refer to contemporary mapping practices, defined as "the emancipatory and subversive effects of mapping practices … that are emerging outside of the cartography traditionally controlled by the state and corporate interests" (Pavlovskaya and St. Martin 2007, 590). Critical cartography can be seen as a form of public cartography. As desktop publishing and geographic information systems (GIS) put mapping into the hands of practically anyone with access to a computer, geospatial technologies have become a means of empowerment, such as research by cancer survivors and volunteers on potential links between environment and breast cancer (McLafferty 2006, 37). A significant aspect of this critical cartography is the taking control of the data and maps, with the proletariat using the data and mapping to their own ends. By taking mapping out of the hands of experts, the projects challenge "orthodoxies" by seeking "to change the world while rewriting accepted ways of theorizing about the world" (Crampton and Krygier 2006, 24; Perkins 2009b, 395). Some scholars, such as Chris Perkins have seen this practice of critical cartography as the "democratization of cartography" (Perkins 2008).

Yet another related term is "counter-mapping," mapping enabled by open-source data creation and sharing, "meaning three things: 1 Anyone is free to obtain them. 2 Anyone is free to change them (i.e., rewrite the code). 3 Anyone is free to distribute them. 'Free' means free to (libre) and often (but not necessarily) financially free (gratis)" (Crampton 2013, 429). Counter-mapping is the antithesis of hegemonic cartography: not just outside of powerful agencies and those controlling geographic knowledge, but open to all, often free for all. Counter-mappings is "genuinely alternative, subversive and emancipatory mapmaking and the degree to which this mapping has effect" (Dodge et al. 2009, 226). As a result of these new technologies, mapping and the creation of new geographic knowledge is occurring in diverse locations.

Recent work in critical cartography has sought to break down the binary between map production and use, questioning the "implicit duality between production and consumption, author and reader, object and subject, design and use, representation and practice" (Del Casino and Hanna 2006, 35). Vincent Del Casino and Stephen Hanna argue for a merging of production and consumption, with no sharp division between where production ends and passive consumption begins. Rather, "Each crease, fold, and tear produces a new rendering, a new possibility, a new (re)presentation, a new moment of production *and* consumption, authoring *and* reading, objectification *and* subjectification, representation *and* practice" (Del Casino and Hanna 2006, 36).[12] So rather than focusing just on map production, scholars need to consider the full range of practices tied with mapping, from their conception, drafting, circulation, use and reuse, all of which depend on contexts and "a mix of creative, tactile, and habitual practices" (Kitchin et al. 2009, 21). These mapping practices are especially important as we consider mapping practices as a significant component of critical cartography or counter-mapping, where it is not just the map that is important but also the act of mapping and the creation of geographic knowledge.

Edney's essay "Cartography without 'progress'" has proven to be the beginning of new work in the history of cartography (Edney 1993). Edney's modes have been picked up and applied, notably in Martin Dodge, Chris Perkins, and Rob Kitchen's "Mapping modes, methods, and moments" where they set out a "manifesto" for map studies (2009). They sum up Edney's version of cartographic history as "a plural and relational network of activities" which they echo, emphasizing the great diversity of current mapping practices, practitioners, and resulting visualizations (Dodge et al. 2009, 221). Dodge, Perkins, and Kitchen call for "new ways of thinking, researching, and creating maps" moving beyond the "tired existing ways of imagining the world" and explore the full potential, "rethinking and remaking your map studies and practice!" (Dodge et al. 2009, 239–40).[13]

Perkins acknowledges, "Networks of practice of map use depend upon relations between many different artefacts, technologies, institutions, environments, abilities, affects, and individuals … we need different ways of approaching mapping and its use, that reflect this complexity… ."

(Perkins 2008, 157). My work on women and geography/cartography in the Progressive Era takes this call and approaches it historically. The democratization of geography and cartography has been occurring for a lot longer than we have previously acknowledged.[14] And while treated as discreet subjects, public geography and cartographic culture together represent a practice of geography outside the halls of power, aspects of a whole. To my knowledge, no studies have yet applied concepts of critical cartography or counter-mapping to historic subjects.[15]

Stitchers of atlases, creators of globes

Scholarship examining women's contributions to cartography began late in the twentieth century, led by the work of Alice Hudson, Judy Tyner, and Mary McMichael Ritzlin (Hudson 1989; Ritzlin 1986, 1989; Tyner 1997, 2001, 2015a, 2015b). Women have long been active in cartography in a wide variety of roles but seldom are these roles addressed by scholars, save in marginal positions, such as map colorists or "stitchers of atlases," although some have been patrons or publishers. With the rise of feminism in geography in the 1970s, women historians of cartography began to delve into women's roles in cartography and uncover the diverse roles they played including "publishers, map sellers, cartographers, drafters, editors, engravers, globemakers, printers, colorists, folders, stitchers, and teachers of map reading and mapmaking, not to mention cartographic historians, map librarians, and patrons of cartography" (Tyner 1997, 46). Much of this research on women and cartography requires piecing together bits and pieces, given little documentation survives capturing these women's contributions and details about their lives, such as their education and how they came to be involved in cartography. Feminist research, drawing from the work of feminist philosopher and theologian Mary Daly, calls for

> ... a different ontology of methods, a process she called 'spooking, sparking and spinning.' *Spooking* means finding those things that 'haunt' forms of knowledge and representation – the absence of women, for example, the deletion of female agency in talking about work done in and around the home, or sexist representations of women... . To *spark* is to restore meaning creatively, making those same absences or silences collide, ironicize, and creatively work together. And *spinning* elaborates those new meanings, moving away from critique of patriarchal approaches, and instead 'hearing forth' new insights. To spin is to stretch, to co-develop our imaginations and thus build and weave new ways of knowing.
>
> Bauchspies and Puig de la Bellacasa 2009, 335[16]

More information is available on famous patrons. Female monarchs, such as Elizabeth I of England and Empress Anna of Russia, ordered surveys of their territories and sponsored mapmakers who dedicated maps and atlases to their sponsors (Ritzlin 1986). Some women came from families of printers and engravers and were raised in the family business, such as Virginia Farrar who continued to update and publish her father's maps after his death (Tyner 1997, 47). Judith Tyner found in Philadelphia in 1820, eleven widows of printers continued their husband's work. Map coloring and atlas binding ("atlas stitching") work employed women until the binding was mechanized in the late 1850s and color printing became possible in the late 1860s (Tyner 1997, 48).

As scholars documented these women cartographers, research progressed from uncovering and naming these cartographers and their contributions, to beginning to consider the ways in which there was a unique women's cartographic culture in the United States (Hudson and Ritzlin 2000).[17] Tyner's work on women's embroidered maps and globes (Tyner 1997, 2001, 2004, 2015a) clearly delineates how girls were taught geography and map-making at boarding schools and seminaries, as early as the 1810s. At Westtown, a Quaker school in Pennsylvania "Westtown school girls may have been the first American globemakers. These were silk, with embroidered graticule and continents, and appeared as early as 1804 – six years before James Wilson, usually considered America's first globemaker, created his" (Tyner 1997, 48). Embroidered maps and globes combined traditional women's skills with knowledge of geography and current events (Tyner 2001, 37). After the 1840s, emphasis in women's education shifted from embroidered maps and globes to "elegant and detailed paper maps" (Tyner 2001, 41). While these woman-made maps and globes may not have significantly advanced our geographic knowledge or represent an advancement in cartographic technique, they reflect the cartographic culture of the time, and even suggest there was a separate women's cartographic culture, created by women, taught to other women, and used to create their own world-views. From Penny Richards' work on Southern women's use of geography and cartographic terms and imagery in their writings, it is clear the geographic education women had as young women led to a lifelong use of geographic and cartographic concepts.

Just as scholars can delineate the "family tree" of cartography so too can we delineate networks of women cartographers, largely independent of the male-dominated academically trained network.[18] This network runs parallel to the public, academic male-dominated cartographic network that has most often been studied during the Progressive Era (for example, see Monmonier and Puhl 2000). This women's parallel network represents an alternative world of geographic knowledge, largely outside of academia, largely undocumented, that created and circulated geographic knowledge and maps for their own ends.

Scholars have long called for work illuminating women's contributions to the history of geography and history of maps and cartography (Huffman 1997; Rose 1993). Like much of the work in the history of geography and in cartography, these discourses are largely separate, yet echo each other in the need to not only recover the presence of women but also to consider feminine practices of these fields.

In Gillian Rose's ground-breaking *Feminism & Geography: The Limits of Geographical Knowledge*, she examines the masculinization of geographical knowledge. Given men's domination of the field, women are marginalized on multiple levels, with topics connected to women and of interest to women not being researched, and with women finding it difficult to find a place for themselves within the discipline (Rose 1993, 2). Rose goes on to consider "the gender of geography" and its implications for feminism. She concludes by suggesting feminism offers an alternative space, a paradoxical space built not on conceptions of Same and Other, that acknowledges difference, and that "can challenge the exclusions of masculinist geography" (Rose 1993, 141). Geography is often constructed as "solid," "science," and "factual" but in reality, "the grounds of its knowledge are unstable, shifting, uncertain and, above all, contested" and there are "other possibilities, other sorts of geographies, with different compulsions, desires and effects, complement and contest each other" (Rose 1993, 160).

Nikolas Huffman, picking up Rose's work and explicitly engaging with cartography, called for research that would:

> ... ground an explicitly feminist cartographic practice that better addresses the concerns of feminist geographical, political, and epistemological theory and serves to recover the power and pleasure of maps to serve feminist interests by highlighting the role of vision and the gendered body in the production of cartographic knowledge.
>
> Huffman 1997, 255

Huffman acknowledges the historically masculinist gaze of maps, in terms of their content, their use, and even the association of landscape with the feminine, to be known and taken. He calls for consideration of how women "see" landscape and map, given differences in gaze and in interests. Chris Perkins echoes Huffman in acknowledging the possible differences in gazes, suggesting "A more hybrid vision becomes possible in which power may be subverted." (Perkins 2009b, 395).

Recent work suggests this "hybrid vision" is being practiced through the combined efforts of feminist geographers and critical cartography. Feminist geographers have engaged in research and/or activism involving geospatial technologies and marginalized peoples, particularly women (Kwan 2007; McLafferty 2006; Pavlovskaya and St. Martin 2007). They argue mapping technologies, such as geographic information systems, while tied to a masculinist perspective, are not inherently masculine, that some aspects are strongly feminist, and that it can be used to "create new ways of looking" (Pavlovskaya and St. Martin 2007, 592). In particular, scholars call on "the development of GT practices that help create a less violent and more just world" (Kwan 2007, 30).

In history of geography scholarship, Avril Maddrell's *Complex Locations: Women's Geographical Work in the UK 1850–1970* (2009) represents exceptional recent work illuminating womens' active roles in geography. Maddrell, using archival research and interviews, delineates women geographers' contributions to the development of UK geography on a wide variety of levels and locations: "They have complex locations in relations to one another and to the institutions and discourses of geographic thought and practice… ." (Maddrell 2009, 3). In her conclusion Maddrell writes: "The sheer volume of women's geography work discussed in this book challenges those disciplinary histories which omit or minimize the presence of women," and she goes on to call for a "rescaling of the map" of the historiography of the discipline to include more of the "small stories" and "minor figures" (Maddrell 2009, 337).

In the history of cartography, Will van den Hoonaard's *Map Worlds: History of Women in Cartography* (2013) ambitiously examines women's contributions to cartography, from the thirteenth century to the present day, employing a range of information, from historical to contemporary. van den Hoonaard's approach is global, not restricted to any particular country, but is somewhat narrowed by his focus on those who consider themselves "cartographers," although he takes a broader approach in historical chapters (van den Hoonaard 2013, 17). His most significant contribution is the interviews he conducted with contemporary women cartographers, with most working within traditional sites of cartography.[19] Through his interviews, a sense of the "small stories" comes through, but not the practice of geography outside of the traditional venues.

Recent research by Sarah Radcliffe has explored the concept of "subaltern cartographies," a practice of cartography where subaltern subjects create a "third space" through their use of map-making, a means of claiming space and generating knowledge (Radcliffe 2011). Subaltern refers to a person holding a subordinate position or a person of inferior station or rank (McEwan 2009, 59). Work on subaltern resistance to hegemony has come from the work of theorists Antonio Gramsci and Homi Bhabha. Subaltern cartography is, in essence, map-making as a form of resistance to colonial/ postcolonial nation-states, countering "official" national maps with maps reflecting a hybrid space of indigenous AND national (Radcliffe 2011). Much of the work on subaltern cartographies has focused on contemporary counter-mapping by indigenous peoples in Latin America and by subjugated peoples in Asia and Africa.

A more interesting concept for me is "subaltern counter-publics" proposed by political scientist Nancy Fraser. In "Rethinking the Public Sphere," Fraser introduces the term, writing

> members of subordinated social groups—women, workers, peoples of color, and gays and lesbians—have repeatedly found it advantageous to constitute alternative publics … parallel discursive arenas where

members of subordinate social groups invent and circulate counterdiscourses, which in turn permit them to formulate oppositional interpretations of their identities, interests, and needs.

Fraser 1997, 81

These "counterdiscourses" which they circulate as a member of the public
to other members of the public allows them to get their message out into
wider spaces (Fraser 1997, 82) This view has been expressed in slightly different terms by Withers who in his work on Victorian England found "several publics to have engaged with geography. There were different audiences
for different sorts of geography and different versions of credibility and
purpose being articulated" (Withers 1999, 69). These counter-public geographies and cartographies can be a means of "linking multiple spaces and
practices through the network form using bonds of affinity, without positing
overarching hegemony or commensurability" (Firth 2014, 160).

Firth's use of the phrase "bonds of affinity" evokes work on affinity politics, the coming together of activist groups for collaboration on common
goals, resulting in unexpected coalitions and partnerships (Larsen and
Johnson 2012). For such a collaboration to take place, groups generally
share certain ethical commitments (Day 2005, 9, 16 and 18). These affinity
politics can result in the creation of affinity spaces, such as Temporary Autonomous Zones (TAZ) or social centers (Day 2005, 35–40). These affinity
spaces bring activists together in a common space, linked by broad common goals. In spaces either squatted or rented, people/groups/causes come
together and "represent an attempt to open up pockets of space that are
dedicated to 'people rather than profit.' These are spaces where people can
experiment, relax, and become involved in a plethora of activities based on
cooperative principles at a grassroots level" (Pusey 2010, 177). In this space,
they are free to pursue their own agendas in the nurturing environment of
shared space/shared ethics and mutual aid. And they often develop out of
a particular "cycle of struggle" (Pusey 2010, 179). Hodkinson and Chatterton identify some of the commonalities found in social centers, including
an emphasis on diversity and difference, the reclamation of private space,
and their function as an activist hub – a location where they can interact,
network, organize (2006, 310). In such spaces, practices of public geographies and cartographies can be taught and learned, linking pedagogy to
both practice and social change, and potentially "bringing new worlds into
being" (Firth 2014, 161).

In examining practices of geography and cartography, it is important
to consider not just the practice and the artifacts of geography/cartog-
raphy but also the affinity networks and the locations where the women
practiced and shared their information. Evidence suggests women of the
Progressive Era did share their methods and encouraged others to conduct studies, create geographic knowledge, and present their findings (see
Chapters 3, 4 and 5).

A women's geographic/cartographic culture of the Progressive Era

In this book, I will be exploring the geographic, cartographic culture of Progressive Era American women, examining the ways women used, produced, employed, wielded, and even embodied maps as part of the work.

Huffman suggests contemporary women's mapping draw on traditional mapping while subverting traditional masculinist representations of space:

> Subversive feminist maps rely on the familiar elements of Western maps; they also employ representations of space that undermine the assumption of an objective and transparently knowable space. Similarly, by introducing women's experiences of space into cartographic images, subversive feminist maps challenge cartography's traditional masculinism.
>
> Huffman 1997, 270

While Huffman was commenting on a contemporary "subversive feminist map" practice, he could have been writing about the Progressive Era. During the Era, scholars such as Frederick Jackson Turner and Halford Mackinder worked at the large scale, on continental, even global levels: "maps were imagined as both products and potential harbingers of geopolitical change" (Heffernan 2002, 208). Progressive Era women's mapping draw on traditional mapping elements while focusing on subjects important to them. Women were busy working on mundane mapping, of playgrounds, of sanitation, of housing conditions, but eventually nationally on suffrage, temperance, anti-lynching, and even globally after 1920 with international peace efforts. They focused on maps as products AND as harbingers of change. While they created and used maps, the subjects they were mapping were a little different than the hegemonic geography/cartography being practiced, as Gillian Rose writes, "women have different spaces, mapped on different grids than men" (Rose 1991, 160).

In approaching women, geography and cartography, I am particularly taken with Donna Haraway's metaphor of cat's cradle, especially as I have been playing with Edney's rope-like schematic. Haraway in her essay "A Game of Cat's Cradle: Science Studies, Feminist Theory, Cultural Studies" considers how scholars might approach issues surrounding nature and technoscience using antiracist feminist theory and cultural studies (Haraway 1994). In particular, how can we approach this in ways that avoid describing technoscience as it has, "relentlessly as an array of interlocking agonistic fields, where practice is modeled as military combat, sexual domination, security maintenance, and market strategy?" (Haraway 1994, 60–1). Rather than using game theory, as has been the case in technoscience, she uses the metaphor of the low-tech children's game cat's cradle, beginning with two "strands" – feminist, multicultural, antiracist technoscience projects and materialized refiguration (reconfiguring what counts as knowledge in the interests of reconstituting the generative forces of embodiment) (Haraway 1994, 62). To this, she adds "colored fibers" of analysis and storytelling, both of which are extremely complicated given the diverse fields Haraway draws on and the complex subjects she is exploring. From these fibers she creates a tangled figure she terms "antiracist multicultural feminist studies of technoscience—i.e., a practice of critical theory as cat's cradle games" (Haraway 1994, 66). She continues, while the game can be played by one, with a large number of patterns, with several pairs of hands involved even more complex patterns can be developed (Haraway 1994, 69). It is not game for "winning," but rather the creation of intriguing patterns. In considering how knowledge is created,

> Tracing networks and configuring agencies/actors/actants in antiracist feminist multicultural studies of technoscience might lead us to places different from those reached by tracing actors and actants through networks in yet another war game. I prefer cat's cradle as an actor-network theory.
>
> Haraway 1994, 71

As do I.

Haraway's approach has much in common with ANT and more recent work done by Bruno Latour, focusing on the intimate connections linking objects, people, and places into intricate networks (Latour 1999). With Latour's Actor-Network Theory, theoretically scholars are "following the actors in different networks, and tracing the inscriptions they leave behind" (Perkins 2008, 152). Given the time period I am working in, these actors are no longer available for interviews or participant–observation work. I am left to draw on texts and visual materials (maps, photographs) to tease out the women's creation and use of cartography. These texts and visual materials are termed in ANT as "immutable mobiles" that is, "information in textual or pictorial form that circulates geographically (it is 'mobile') while at the same time retaining its meaning (it is 'immutable')" (Barnes 2006, 153). With this immutable format, knowledge is stabilized for its circulation, such as geographic information in the form of reports and maps. Geographers, such as Trevor Barnes, have used ANT to examine the spread of geographical concepts (Barnes 2006). Despite the concept, immutable mobiles seldom are truly immutable. Users take the "mobiles" and do what they wish. Further, texts interact with other texts (Perkins 2009b, 396). On one hand, immutable mobiles work to reify dominant ideologies, yet users have the ability to resist these ideologies and use them to their own ends (Del Casino and Hanna 2006, 42).

In critical human geography as well as in recent work on social aspects of cartography, there has been a move since the 1980s towards considering mapping process, drawing particularly from nonrepresentational theories that pay particular attention to practices and performance (Kwan 2007, 23).

The ANT concept of assemblage, referring to an identifiable entity composed of actors and actants that have become connected in some way, has been applied to mapping:

> The practice of mapping, whether it is scientific-technical, historical-hermeneutic, or critical depends upon the context, but also influences the context. Different assemblages interact: the Western mapping tradition adapts to others it encounters, and so adopts elements of these traditions, approaches, and philosophies. Within that tradition, <u>networks or webs of practice enact different mapping philosophies</u>… . Mapping itself, and philosophies of mapping, have an agency of their own, capable of enacting change.
>
> Perkins 2009b, 388–9[20]

As I approach Progressive Era women's creation of geographic knowledge and maps, it is not just the immutable mobiles that are significant, it is also the assemblages, the practices of mapping and their networks, I will be examining.

Given the complexity of the material at hand – the wide variety of groups, the various locations, the different work they did—there are many ways this far-reaching topic could be explored. The advantage of the "cat's cradle" approach is its acknowledgement that the resulting pattern is just one way of thinking about the issue at hand, avoiding the criticism of ANT that by focusing on networks, it leaves no spaces to exist outside the network, resulting in a flat, centered view (Hetherington and Law 2000).

Cat's cradle is a very appealing metaphor to me in approaching this work on women, geography and cartography. As the winding route of this chapter demonstrates: not one approach can take everything into account. There needs to be a complex web: of work on women's history and feminist theory in thinking about women's lives and experiences; of work on the history of geography and cartography; of recent work considering the practice of geography and cartography outside of academics and government; and so on. To focus on one aspect would neglect an important thread or multiple threads. This work represents just one possible figure of women, geography, and cartography in the Progressive Era.

And so, I take up the threads first arrayed by Matthew Edney … who was inspired by a diagram created by Penny Richards for her master's thesis. Edney, Richards, and myself all attended graduate school in Geography at the University of Wisconsin-Madison and were all connected with the History of Cartography Project directed by the historian of cartography David Woodward and the historical geographer J. Brian Harley. Richards' thesis diagram inspired Edney's schematic, which he acknowledges in a recent reflection piece on his essay "Cartography without progress," now viewed as a "classic" in cartography (Edney 2011, 334). And scholarship by Edney and Richard are essential for me in approaching this research. I now pick up

Edney's schematic and add new threads to the configuration, creating a version that makes a space for women's practice of the history of cartography and geography.

I begin in Chapter 2 by examining the increase in women's use of maps for wayfinding brought about by the development of the bicycle and later the automobile. Using maps for navigating our way through the world is not a new concept. It is likely some of the earliest maps ever constructed were scratched into dirt or the ashes of a fire as humans discussed how to get from one place to another. Wayfinding maps are some of the oldest extant maps known to modern humans (Akerman 2007, 21–3). Given their utilitarianess, such maps are often perceived as straightforward and objective. But in the 1890s when a boom in bicycling led to greater need for maps for this purpose, women's realms were largely circumscribed around home and family and so their mobility beyond the domestic sphere was politically fraught, with questions being raised about the implications of mobility. Bicycling and then automobiling led to women-specific, and even women-generated, directions and mapping. The bicycle and all its accoutrements entered American women's lives and began to transform their understanding of the world around them. As women rode their bicycles, they experienced landscape in new ways, they went farther (spatially and socially) as a result, and they became more aware of their geography. The automobile extended this even further. The technological developments of the bicycle and the automobile changed *how* Americans moved through the landscape, what they knew about the land, *and* even the landscape itself.

In Chapters 3 and 4, I consider "woman's work." The phrase "woman's work" was associated with social improvement activities carried out outside the home. Chapter 3 examines the "woman's work" carried out by missionaries and their use of geography and cartography in the work. A significant aspect of missionary work was mobilizing allies to support their work, spiritually as well as financially. Through their writings, women missionaries shaped the geographical imaginations of their audiences, both with geographical descriptions and with spatial information, and promoted cartographic literacy. While much was basic geography, it also informed their audiences on gender specific issues – women's rights in marriage, infanticide, suttee, foot binding—bridging worlds as women strove to help other women and children.

In Chapter 4, I continue to explore "woman's work," specifically social settlements and their use of studies in their social activism, largely to improve their own communities as part of what is termed "municipal housekeeping." Beginning also in the 1890s, women sought a more active presence outside the home, attempting to address what they saw as social issues that had not been tackled, such as sanitation, living conditions, disease, and food issues. Through a variety of organizations and structures, they conducted studies of their communities, created reports and maps, and lobbied to remedy these problems. Using examples from settlement work, specifically Hull-House in Chicago, and organizations such as the Women's Christian Temperance Union, I consider their creation of geographic knowledge through studies and maps. Moreover, they encouraged other women to engage in such practices, providing directions and encouragement, in practices of critical cartography and counter-mapping.

I explore a different aspect of social activism in Chapter 5, examining how women used geographic knowledge and maps as part of their political activism rhetoric to sway public interest on their causes. I focus largely on the use of a persuasive map by the suffrage movement, used extensively around the country from approximately 1906 until they received the vote in 1919. In the case of the suffrage map, the National American Woman Suffrage Association took a relatively straightforward informational map and crafted it into a propaganda map that was plastered across the country, to the point where suffragists became the map (performance cartography). This is not the only example of persuasive map use by American women in the Progressive Era but represents a spectacular, large-scale use of maps by American women and suggests the sophisticated level of geographic/cartographic practice by American women. I also briefly discuss the maps used and created by women in efforts to end the practice of lynching.

Finally, in Chapter 6, I conclude my exploration of geographic/cartographic practice by American women in the Progressive Era and my game of cat's cradle, reflect on my completed figure, and suggest some possible expansions of the game.

Progressive Era women in their practices of geography and cartography were actively creating knowledge (Crampton 2013, 425). Moreover, they were using Americans' geographic/cartographic culture to their advantage. They were well aware of how geography and maps worked: how they "speak," what they embody, what they represent, and how they represent. By creating geographic knowledge and using formal conventions to present their information, they were addressing an audience who was accustomed to formal cartography and depending on them to read and interpret their work with the same considerations.

Nikolas Huffman, writing in 1997, called for "an explicitly feminist cartographic practice" that would assist in recovering "the power and the pleasure of maps to serve feminist interests" (Huffman 1997, 255). I believe a version of this practice already existed in the Progressive Era: they were already experiencing the power and the pleasure of maps. The knowledge Progressive Era women were creating reflects their situated knowledge and interests and are tied to "complex networks of knowledge, discourse, media forms, technologies and networks of power and patronage" (Kitchin et al. 2013, 494). American women were making claims, taking control of space as well as knowledge, creating their own representations, and making statements about what they knew about the world.

We will begin our exploration of Progressive Era women's geography with navigation, finding their way in the world.

Notes

1 Morse dedicated another of his geographies – *Geography Made Easy* – "To the Young Masters and Misses Throughout the United States" (Tolley 2003, 13). *Geography Made Easy* was first published in 1784.

2 Withers comments that jigsaw puzzles were an important education medium through which "geography, political awareness, and politeness were all taught" (Withers 2011, 47).

3 Kelley also briefly discusses Charlotte Forten, a free African-American who was sent to Salem, Massachusetts in 1854 to complete her education, where her studies included "modern geography" (Kelley 1996, 409).

4 Geography has a renaissance at European universities in the nineteenth century. It had a "powerful presence" at German universities. In France, "a dozen new chairs [in geography] were established during the 1880s and the 1890s." In Britain, a full-time university post in geography was not created until 1887 when one was awarded to Halford Mackinder (Heffernan 2009, 12).

5 Edney's modes are echoed in Charles Withers work on "Geography's Narratives and Intellectual History" where he structures his approach around geographic practices such as writing and reading, mapping, and depicting, exploring and trusting, experimenting and gesturing and conversing. Withers writes: "My concern is to show how, at different times, the *art* of doing geography (including within that term the science in and of geography), has always also been, in different ways, the *act* of doing geography" (Withers 2011, 40). Emphasis is from Withers.

6 Edney has acknowledged drawing inspiration from a visualization Penny Richards created for her master's thesis on the interplay of spatial information and perception regarding northern Pennsylvania in the eighteenth century (Edney 2011, 334).

7 The painting hangs today over the west staircase of the Senate wing of the U.S. Capitol (Schulten 2012, 119).

8 The American Geographical Society at that time was seen as dominated by "elderly amateurs" (Lutz and Collins 1993, 20).

9 Emphasis is from Withers (1999).

10 Catherine Delano Smith, a contributor to Volume One of the series, used the phrase "cartographic culture" in a 1990 book review and then called for work on map ownership and map use in a 1995 article on map ownership (Smith 1990, 1995). My thanks to Matthew Edney for his thoughts and research on the origins of the term "cartographic culture." Personal correspondence between Dando and Edney, 14th August 2014.

11 Emphasis is from Edney (2011).

12 Emphasis is Del Casino and Hanna's.

13 Emphasis is Dodge et al's.

14 I am not suggesting the critical cartography of the past is exactly the same as the critical cartography of today. Today's critical cartographers are "aware of gender and other dimensions of power (e.g. class, race, heterosexuality) and advocate progression politics that destabilize these power hierarchies" (Pavlovskaya and St. Martin 2007, 592). In the case of the suffrage movement, while aware of power differences and generally advocating progressive politics, a conscious political decision was made to excluding any race other than white in order to achieve the goal of suffrage for the majority of their constituents.

15 However, former slave John Washington's map of Fredericksburg VA can been interpreted as an example of subaltern cartography. See Hanna (2012).

16 Emphasis is from the original.

17 British schoolgirls created embroidered maps as early as the 1770s before the idea spread to the United States. Embroidered maps waned after 1840 (Tyner 2001, 37).

18 See Matthew Edney's "Putting 'Cartography' into the History of Cartography" (2005) and its accompanying "academic family tree" created by Henry Castner.
19 van den Hoonaard's work, while interesting, has its critics. See Judith Tyner's review in *The AAG Review of Books* (2014).
20 Emphasis added.

Bibliography

Akerman, J. 2007. Finding our way. In *Maps: Finding Our Place in the World*, eds. J. Akerman and R. Karrow, pp. 19–63. Chicago, IL: The University of Chicago Press.

Barnes, T. 2006. Geographic intelligence: American geographers and research and analysis in the Office of Strategic Services 1941–1945. *Journal of Historical Geography* 32: 149–68.

Bauchspies, W. and M. Puig de la Bellacasa. 2009. Feminist science and technology studies: a patchwork of moving subjectivities. An interview with Geoffrey Bowker, Sandra Harding, Anne Marie Mol, Susan Leigh Star and Banu Subramaniam. *Subjectivity* 28: 334–44.

Brown, V. 2004. *The Education of Jane Addams*. Philadelphia, PA: University of Pennsylvania Press.

Brückner, M. 1999. Lessons in geography: maps, spellers, and other grammars of nationalism in the Early Republic. *American Quarterly* 51/2: 311–43.

Cosgrove, D. 2008. *Geography and Vision: Seeing, Imagining and Representing the World*. London/New York: I.B. Tauris & Co. Ltd.

Crampton, J. 2013. Mappings. In *Wiley-Blackwell Companion to Cultural Geography*, eds. N. Johnson, R. Schein and J. Winders, pp. 423–36. New York/London: John Wiley & Sons, Ltd.

Crampton, J. and J. Krygier. 2006. An introduction to critical cartography. *Acme: An International E-Journal for Critical Geographies* 4, 1: 11–33.

Dando, C. 2002. *Going Places?: Gender and Map Use in 20th Century Road Map Art*. The Newberry Library Slide Set Number 33. Chicago, IL: The Henry Dunlop Smith Center for the History of Cartography, The Newberry Library.

———. 2003. *"Happy Motoring!: selling maps and geographic information to 20th century American women,"* paper presented at the International Conference on the History of Cartography, Boston, MA and Portland, ME, June 2003.

Day, R. 2005. *Gramsci Is Dead: Anarchist Currents in the Newest Social Movements*. London and Ann Arbor: Pluto Press and Toronto: Between the Lines.

Del Casino, V. and S. Hanna. 2006. Beyond the "binaries": a methodological intervention for interrogating maps as representational practices. *Acme: An International E-Journal for Critical Geographies* 4, 1: 34–56.

Dillon, D. (undated), 'Agnes Sinclair Holbrook,' http://tigger.uic.edu/htbin/cgiwrap/bin/urbanexp/main.cgi?file=new/show_doc.ptt&doc=280&chap=39. Accessed 12 December 2013.

Dodge, M., C. Perkins, and R. Kitchin. 2009. Mapping modes, methods and moments. In *Rethinking Maps*, eds. M. Dodge, C. Perkins, and R. Kitchin, pp. 220–43. London: Routledge.

Edney, M. H. 1993. Cartography without 'progress': reinterpreting the nature and historical development of mapmaking. *Cartographica* 30, 2 &3: 54–68.

———. 1994. Cartographic culture and nationalism in the early United States: Benjamin Vaughan and the choice for a prime meridian, 1811. *Journal of Historical Geography* 20, 4: 384–95.

———. 1997. *Mapping an Empire: The Geographical Construction of British India, 1765–1843.* Chicago: University of Chicago Press.

———. 2005. Putting 'cartography' into the history of cartography: Arthur H. Robinson, David Woodward, and the creation of a discipline. *Cartographic Perspectives* 51, Spring: 14–29.

———. 2011. Reflection essay: progress and the nature of 'cartography.' In *Classics in Cartography: Reflections on Influential Articles from Cartographica*, ed. M. Dodge, pp. 331–42. London: John Wiley & Sons, Ltd.

Firth, R. 2014. Critical cartography as anarchist pedagogy? Ideas for praxis inspired by the 56a infoshop map archive. *Interface: a journal for and about social movements* 6, 1: 156–84.

Flanagan, M. 1990. Gender and urban political reform: the City Club and the Woman's City Club of Chicago in the Progressive Era. *The American Historical Review* 95, 4: 1032–50.

Fraser, N. 1997. *Justice Interruptus: Critical Reflections on the 'Postsocialist' Condition.* New York: Routledge.

Friendly, M. and G. Palsky. 2007. Visualizing nature and society. In *Maps: Finding Our Place in the World*, eds. J. Akerman and R. Karrow, pp. 207–53. Chicago, IL: The University of Chicago Press.

Hanna, S. 2012. Cartographic memories of slavery and freedom: examining John Washington's map and mapping of Fredericksburg, Virginia. *Cartographica* 47, 1: 50–63.

Haraway, D. 1994. A game of cat's cradle: science studies, feminist theory, cultural studies. *Configurations* 2, 1: 59–71.

Harley, J.B. 1988a. Silences and secrecy: the hidden agenda of cartography in early modern Europe. *Imago Mundi* 40: 57–76.

———. 1988b. Maps, knowledge, and power. In *The Iconography of Landscape: Essays on the Symbolic Representation, Design, and Use of Past Environments*, ed. D. Cosgrove and S. Daniels, pp. 277–312. Cambridge: Cambridge University Press.

Heffernan, M. 2002. The politics of the map in the early twentieth century. *Cartography and Geographic Information Science* 29, 3: 207–26.

———. 2009. Histories of geography. In *Key Concepts in Geography*, eds. N. Clifford, S. Holloway, S. Rice and G. Valentine, pp. 3–22. London/Thousand Oaks, CA: SAGE Publications, Ltd.

Hetherington, K. and J. Law. 2000. After network. *Environment and Planning D: Society and Space* 18: 127–32.

Hodkinson, S. and P. Chatterton. 2006. Autonomy in the city? Reflections of the social centres movement in the UK. *City* 10, 2: 305–15.

Hudson, A. 1989. Pre-twentieth century women mapmakers. *Meridian* 1: 29–32.

Hudson, A. and M. Ritzlin. 2000. Checklist of pre-twentieth-century women in cartography. *Cartographica* 37, 3: 9–24.

Huffman, N. 1997. Charting the other maps: cartography and visual methods in feminist research. In *Thresholds in Feminist Geography: Difference, Methodology, Representation*, eds. J. Jones III, H. Nast and S. Roberts, pp. 255–283. Lanham, MD: Rowman & Littlefield Publishers, Inc.

Jackson, S. 2000. *Lines of Activity: Performance, Historiography, Hull-House Do-mesticity.* Ann Arbor, MI: University of Michigan Press.

James, P. and G. Martin. 1981. *All Possible Worlds: A History of Geographical Ideas.* New York: John Wiley & Sons.

Jerabek, E. 1943. Some sources for Northwest history: early geography textbooks. *Minnesota History* 24, 3: 229–33.

Kelley, M. 1996. Reading women/women reading: the making of learned women in antebellum America. *The Journal of American History* 83, 2: 401–24.

Kitchin, R., J. Gleeson, and M. Dodge. 2013. Unfolding mapping practices: a new epistemology for cartography. *Transactions of the Institute for British Geographers* NS 38: 480–96.

Kitchin, R., C. Perkins, and M. Dodge. 2009. Thinking about maps. In *Rethinking Maps*, eds. M. Dodge, C. Perkins and R. Kitchin, pp. 1–25. London: Routledge.

Kwan, M. 2007. Affecting geospatial technologies: toward a feminist politics of emotion. *The Professional Geographer* 59, 1: 22–34.

Larsen, S. and J. Johnson. 2012. Towards an open sense of place: phenomenology, affinity and the question of being. *Annals of the Association of American Geographers* 102, 3: 632–46.

Latour, B. 1999. Science's blood flow. In *Pandora's Hope: Essays on the Reality of Science Studies.* Cambridge, MA/London: Harvard University Press.

Livingstone, D. 2000. Putting geography in its place. *Geographical Research* 38, 1: 1–9.

Lorimer, H. 2003. Telling small stories: spaces of knowledge and the practice of geography. *Transactions of the Institute of British Geographers* NS 28: 197–217.

Lutz, C. and J. Collins. 1993. *Reading National Geographic.* Chicago, IL: University of Chicago Press.

Maddrell, A. 2009. *Complex Locations: Women's Geographical Work in the UK 1850–1970.* Chichester: Wiley-Blackwell.

McCammon, H., K. Campbell, E. Granberg, and C. Mowery. 2001. How movements win: gendered opportunity structures and U.S. women's suffrage movements, 1866 to 1919. *American Sociological Review* 66, 1: 49–70.

McEwan, C. 2009. Social and cultural geography: subaltern. In *The International Encyclopedia of Human Geography*, eds. R. Kitchin, & N. Thrift, pp. 59–64. Amsterdam: Elsevier.

McLafferty, S. 2006. Women and GIS: geospatial technologies and feminist geographies. *Cartographica* 40, 4: 37–45.

Monmonier, M. and E. Puhl. 2000. The way cartography was: a snapshot of mapping and map use in 1900. *Historical Geography* 28: 157–78.

Morin, K. 2011. *Civic Discipline: Geography in America, 1860–1890.* Farnham, UK: Ashgate.

Pavlovskaya, M. and K. St. Martin. 2007. Feminism and geographic information systems: from a missing object to a mapping subject. *Geography Compass* 1/3: 583–606.

Perkins, C. 2008. Cultures of map use. *The Cartographic Journal* 45, 2: 150–58.

———. 2009a. Performative and embodied mapping. In *International Encyclopedia of Human Geography*, eds. R. Kitchin and N. Thrift, pp. 126–32. Amsterdam: Elsevier.

———. 2009b. Philosophy and mapping. In *International Encyclopedia of Human Geography*, eds. R. Kitchin and N. Thrift, pp. 385–97. Amsterdam: Elsevier.

Pusey, A. 2010. Social centres and the new cooperativism of the Common. *Affinities: A Journal of Radical Theory, Culture and Action* 4, 1: 176–98.

Radcliffe, S. 2011. Third space, abstract space, and coloniality: national and subaltern cartography in Ecuador. In *Postcolonial Spaces: The Politics of Place in Contemporary Culture*, eds. A. Terson and S. Upstone, pp. 129–45. Basingstoke: Palgrave Macmillan.

Richards, P. 2004. 'Could I but mark out my own map of life': educated women embracing cartography in the nineteenth-century American South. *Cartographica* 39, 3: 1–17.

Ritzlin, M. 1986. The role of women in the development of cartography. *AB Bookman's Weekly* June 9, 1986: 2709–13.

———. 1989. Women's contributions to North American cartography: four profiles. *Meridian* 2: 5–16.

Rose, G. 1993. *Feminism & Geography: The Limits of Geographical Knowledge.* Minneapolis, MN: University of Minnesota Press.

———. 1991. On being ambivalent: women and feminism in geography. In *New Words, New Worlds: Reconceptualizing Social and Cultural Geography*, ed. C. Philo, pp. 156-63. Aberystwyth [Wales]: Cambrian Printers.

Rynbrandt, L. 1997. The 'Ladies of the Club' and Caroline Bartlett Crane: affiliation and alienation in Progressive social reform. *Gender & Society* 11, 2: 20014.

Schulten, S. 2001. *The Geographical Imagination in America: 1880–1950.* Chicago, IL: University of Chicago Press.

———. 2012. *Mapping the Nation: History and Cartography in Nineteenth-Century America.* Chicago, IL: University of Chicago Press.

Smith, C. 1990. Review of *La Città de Napoli tra Vedutismo e Cartografica. Imago Mundi* 42: 139.

———. 1995. Map ownership in sixteenth-century Cambridge: the evidence of probate inventories. *Imago Mundi* 47: 67–93.

Smith, J. 2013. Review essay – geography in public and public geography: past, present and future. *The Geographical Journal* 179, 2: 188–92.

Stebner, E. 1997. *The Women of Hull-House: A Study in Spirituality, Vocation, and Friendship.* Albany, NY: State University of New York Press.

Tolley, K. 2003. *The Science Education of American Girls: A Historical Perspective.* New York: RoutledgeFalmer.

Townsend, L. 1990. The gender effect: the early curricula of Beloit College and Rockford Female Seminary. *The History of Higher Education Annual* 10: 71–92.

Tyner, J. 1997. The hidden cartographers: women in mapmaking. *Mercator's World* 2, 6: 46–51.

———. 2001. Following the thread: the origins and diffusion of embroidered maps. *Mercator's World* 6, 2: 36–41.

———. 2004. Stitching the world: Westtown School's Embroidered Globes. *Piecework* 12, 5: 28–31.

———. 2014. Review of *Map Worlds: A History of Women in Cartography. The AAG Review of Books* 2, 1: 22–24.

———. 2015a. *Stitching the World: Embroidered Maps and Women's Geographical Education.* London: Routledge.

———. 2015b. Women in cartography. In *History of Cartography, Volume 6 Cartography in the Twentieth Century*, ed. M. Monmonier, pp. 1758–61. Chicago, IL: University of Chicago Press.

van den Hoonaard, W. 2013. *Map Worlds: A History of Women in Cartography.* Waterloo, ON: Wilfrid Laurier University Press.

Withers, C. 1999. Towards a history of geography in the public sphere. *History of Science* 37, 1, 115: 45–78.

———. 2006. Eighteenth-century geography: texts, practices, sites. *Progress in Human Geography* 30, 6, 711–29.

———. 2011. Geography's narratives and intellectual history. In *The SAGE Handbook of Geographical Knowledge*, eds. J. Agnew and D. Livingstone, pp. 39–50. London, UK: Sage.

Wooldridge, S. and W. East. 1951. *The Spirit and Purpose of Geography.* London: Hutchinson.

2   Women, geographic knowledge, and mobility

> After fifteen miles of alternate sand and mud, hills and bogs, and a cold wind blowing, of course, straight in my face, I decided to stop in the next town and spend the rest of the day expressing my opinion about the League map of Iowa, which is a snare and a delusion.
>
> Le Long 1898a, 492

> By this time I had developed a positive mania for getting on the wrong road, so I made inquiries of all the cyclers and old pioneers in Laramie, wrote down their directions, drew maps of the country, and then took the wrong road. It landed me on the banks of an alkali lake, and ended there.
>
> Le Long 1898b, 593

Margaret Valentine Le Long, in her account of bicycling from Chicago to San Francisco in the late 1890s, vented her frustration with geographic information as she made her way across the American heartland. In the top quote above, her League of American Wheelmen map of Iowa was found wanting, not accurately conveying the road conditions (for an example of a League map, see Figure 2.1).[1] The League maps were viewed as the best roads maps available, assembled from information provided by state League members and showing not only the locations of roads and towns, but coded to indicate the types of roads, the conditions of the roads, and the grade. In the bottom quote, Margaret gathered as much information as she could about Wyoming's roads, but that was still not enough and she lost the road. A local man eventually came along and when Le Long asked where the road was, he replied she was standing in the middle of it, acknowledging "it wasn't 'any great shakes of a road, nohow'" and pointing her in the right direction (Le Long 1898b, 593).

Margaret did make it all the way to San Francisco by bicycle, despite her frustrations with the information and infrastructure available to her. This example of a woman critiquing her map and the geographic information available to her, questioning its authority as she navigated her way across the landscape, is striking. It captures the challenges of the early female

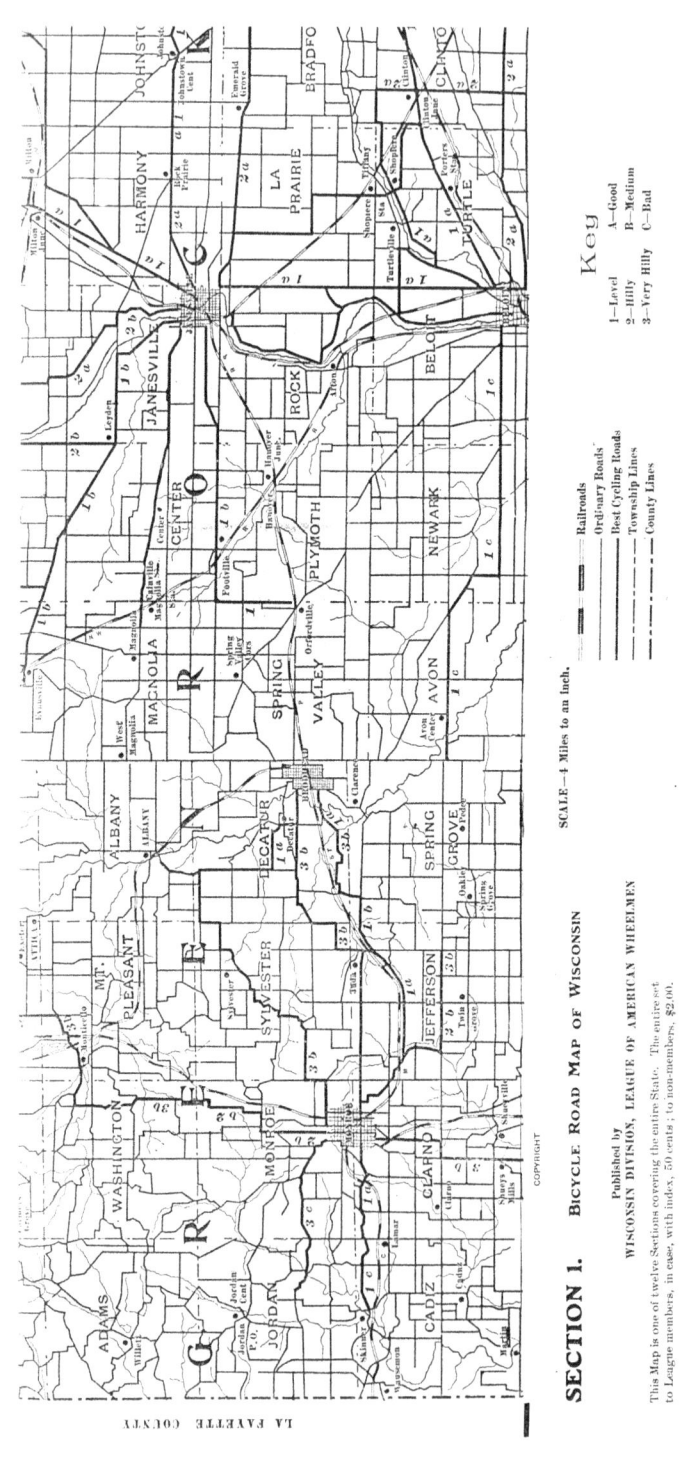

SECTION 1.   BICYCLE ROAD MAP OF WISCONSIN

Published by
WISCONSIN DIVISION, LEAGUE OF AMERICAN WHEELMEN

This Map is one of twelve Sections covering the entire State. The entire set to League members, in case, with index, 50 cents ; to non-members, $2.00.

SCALE—4 Miles to an Inch.

Key

1—Level   A—Good
2—Hilly   B—Medium
3—Very Hilly C—Bad

Railroads
Ordinary Roads
Best Cycling Roads
Township Lines
County Lines

Figure 2.1 Example of a League of American Wheelmen cycling map: Bicycle Road Map of Wisconsin. This section from southwestern Wisconsin illustrates how the League keyed the roads not only in terms of terrain (level, hilly, very hilly), but also in terms of road quality (good, medium, bad). Riders could then plan their routes according to both road conditions and how challenging a ride they wished to undertake. By permission of the Wisconsin Historical Society, Madison, WI (WHi-50449; Image ID 50449).

bicyclists, both the need for precise geographic information but also the need for better infrastructure to ride the bicycles: women contributed to both during the Progressive Era.

To get from one location to another, whether it is near or far, by foot, bicycle, or automobile, requires an understanding of geography; that geography might just be of a block or two, or it might be of a vast territory. We often think of geographic knowledge as static, as fixed – mountains, rivers, boundaries, capitals – but even these are in motion … mountains are constantly moving on their continental plates as they simultaneously are building *and* eroding, rivers flow and can shift their bed in times of high water, boundaries can change through treaty or war, and capitals can rise and fall or be relocated for commercial purposes. Human too have always been on the move. It was a necessity for early humans to move almost constantly to maintain the nutrient supply to keep them alive (hunting-gathering). As we developed new technologies, we moved. The development of sedentary agriculture meant that while we now had stable food supplies and did not need to be on the constant move, yet stable food led to larger populations which in turn led to populations moving to new areas. Geography is, in many ways, about movement, about mobility – our world is constantly changing and we are constantly moving.

Women in the Progressive Era found greater mobility during this time thanks to the technological advances of the age, especially the bicycle and then the automobile.[2] As women became more mobile, they engaged with geography and maps in new ways. Prior to the advances of the bicycle which permitted individual mobility on an unprecedented level, mobility for women was limited to walking, horseback, wagon or carriage, and trains for longer distances. Individual levels of mobility varied greatly for women depending on a number of factors, including class, race, and occupation. Scholars today recognize that while there was definitely a *rhetoric* in the American media that described women and men as occupying "separate spheres," with men associated with the public and women with the private (home), this was not the *reality* for most women (Kerber 1988; Mackintosh and Norcliffe 2007, 154–5; McGaw 1989). Lower and middle-class women were not restricted to the home sphere: many worked outside the home in shops, in factories, or in homes as domestics. Even if not employed outside the home, they might travel to help with the family livelihood, to visit family, to shop, or to take pleasure in a park during a moment of leisure. Their geographic knowledge was likely split between a basic understanding of the world (general knowledge of countries and locations learned in school) and a practical understanding of the paths they needed to take to move from home to work, from home to market, from home to places of worship, from home to the homes of family and friends. Wealthy women traveled if they wished and if their families permitted them, with many traveling to Europe to "finish" their education (Levenstein 1998, 184). Some women toured regularly because of their interests, both within the United States and outside (Morin 2008). Women's geographic knowledge differed given the different landscapes they occupied and the different ways they moved through the landscape.

Both the bicycle and the automobile introduced new methods of mobility that involved new practices and new knowledges, including new geographies. As we move – be it by train, bicycle, or automobile – we are involved in the "crafting of time and space"; activities and practices are involved in the movement, links are made and broken, social and spatial relations are arranged and rearranged (Watts 2008, 712). As we craft time and space, a variety of geographic knowledges are employed, from the physical and cultural geographies of the landscapes we move through to the spatial knowledge needed to get from point A to point B. This chapter explores the new geographical knowledges needed and created by these new mobilities by American women. Not all women were able to take advantage of these new mobilities but many did. These new mobilities involved new practices and new knowledge but also significantly transformed the American landscape, a process that many women were actively involved in.

Girl on a wheel: Women and bicycling

> I began to feel that myself plus the bicycle equaled myself plus the world, upon whose spinning wheel we must all learn to ride, or fall into the sluiceways of oblivion and despair. That which made me succeed with the bicycle was precisely what had gained me a measure of success in life—it was the hardihood of spirit that led me to begin, the persistence of will that held me to my task, and the patience that was willing to begin again when the last stroke had failed. And so I found high moral uses in the bicycle and can commend it as a teacher without pulpit or creed. He who succeeds, or, to be more exact in handing over my experience, she who succeeds in gaining mastery of such an animal as Gladys [the name she gave her bicycle], will gain the mastery of life, and by exactly the same methods and characteristics.
>
> Frances Willard 1895/1991, 32–33

Frances Willard was fifty-three years old when she learned to ride the bicycle in 1893. Raised in Illinois in a middle-class family, Frances attended college in Wisconsin and Illinois and eventually taught at the North Western Female College in Evanston. She became involved with the Woman's Christian Temperance Union (WCTU), a national movement to address the growing problem of alcoholism, and eventually became its national president, a position she held for the last twenty years of her life. Visiting friends in England, she was presented with a bicycle and was determined to learn (Mayo 1991, 10–11). After mastering the bicycle, Willard wrote *A Wheel Within a Wheel* (1895), encouraging others to master the bicycle and linking these skills to other life skills.[3] It became a national bestseller. While emphasizing empowerment of all through the bicycle, Willard was careful to frame cycling as a genteel, womanly activity appropriate for "proper" women (Mackintosh and Norcliffe 2007, 165).

The bicycle was at the center of the transformation in American women's lives in the late nineteenth century. Through print ads, advertising art, articles, and books, a new world for women was being created and promoted. A woman's public space was being fashioned in which they could move, that women controlled and that allowed women to network and nurture their own culture (Kerber 1988, 32). While these geographies were still male dominated and controlled, the marketing and promotion of the bicycle expanded women's presence in American society and geography's presence in women's lives.

American women's adoption of the bicycle

Women's adoption of the bicycle was the result of a complex set of advancements: innovations in technology that produced a bicycle that women could easily ride, progress culturally that opened the door on women's mobility including attitudes about women's social place, about women's clothing, about women's place in public, improvements in production that made the bicycle and its related paraphernalia affordable for a wider number of Americans.

First and foremost were advances in the design of the bicycle that made it easier for women to master the wheel (Figure 2.2). Early versions of the bicycle had been around since the early 1800s in the form of velocipede which had limited adoption by women given its heavy weight, difficulty in riding in skirts, and social taboos (Petty 1997, 113–14). The high wheel or Ordinary was invented in 1874. With its large front wheel and high seat, it was mostly adopted by young men who were undeterred by the very real risks of a dangerous header over its handlebars. Its design made it very difficult for women to master given the dress of the day with long full skirts and whalebone corsets (although there were women who mastered it). In 1876, a tricycle was developed and achieved some popularity with women, but its size, weight, and expense limited its audience (Petty 1997, 117).[4] Frances Willard was an early tricycling enthusiast, her tricycle a gift from Albert Pope, the founder of Columbia bicycles (Willard 1895/1991, 19).

The first commercially successful safety bicycle was produced in 1885, with two equal-sized wheels (the "Rover" created by John Kemp Starley) (Epperson 2000, 310; Wosk 2001, 97). The development of pneumatic tires in 1887 made for a smoother, less jarring ride. A series of ladies' bicycles were produced by a variety of companies in 1887–1889, based on a safety frame with a lowered top tube to allow cycling in skirts. Technological advances made bicycling safer, more comfortable, and eventually reasonably priced, leading to millions trying it, with cycling peaking as a fad in 1895 (Petty 1997, 118). *Munsey's* magazine in 1896 proclaimed that "To men, the bicycle in the beginning was merely a new toy, another machine added to the long list of devices they knew in their work and in their play. To women, it was a steed upon which they rode into a new world" (The World Awheel 1896, 157).

Figure 2.2 Columbia bicycle, advertisement, from *Scientific American* 78, 16 (April 16, 1898): 256.

While the design of the bicycle had evolved into a form that was more conducive to women riding, women's dress still restricted cycling. As more women took to bicycling, women's dress began to adapt to these new activities.[5] Women's rights activists had called for women's dress reforms (termed "rational dress") since the 1850s. One alternative was the "bloomer," a shortened skirt with baggy pantaloons underneath, which Amelia Bloomer promoted in her women's newspaper *The Lily* and wore herself (Macy 2011a, 45–8). But bloomers were only adopted by a few. With the bicycle, women began to adapt their dresses to this new mobility. Some took up the bloomers. Others wore either slightly shorter skirts or divided skirts. Initially, women made their own bicycling costumes or had them made locally. An 1894 ad in *Godey's Lady's Book and Magazine* advertises a pattern for "The

Columbia Bicycle Habit": "Believing that if ladies better understood what to wear awheel, more wheelwomen would engage in the incomparable sport of cycleing, we have had designed for us, by Redfern of New York and London, the accompanying stylish and graceful costume… ."[6] In addition to skirts, another element of women's dress that evolved with cycling was the corset. "Bicycling corsets" were produced that were less restrictive and available commercially (Bogardus 1998; Macy 2011a, 2011b, 44–5; Petty 1997, 128). While the subject of dress we often associate with appearance, there was a practical element to the issue of women's dress while cycling:

> On the excursion a special adaptation of dress is absolutely necessary, for skirts, while they have not hindered women from climbing to the topmost branches of higher education, may prove fatal in down-hill coasting. …
>
> Merington 1895, 703

Appearance was important but so was the practical need for a costume that would not result in a serious accident.

These changes to women's dress were significant advancements in the lives of American women. They were also one of the elements of bicycling that was most hotly debated: it was not just a matter of women moving freely in the world, it was also *how* they moved through the world.[7] The matter of women's dress while bicycling was publicly debated, eventually parodied in cartoons. Nearly all guides, articles, and other publications on women bicycling addressed dress, advocating cycling costumes that maintained a lady-like appearance while facilitating movement. For example, Margaret Le Long towards the end of her account from Chicago to San Francisco writes:

> To my cycling sisters I wish to say that I rode a drop-frame wheel, and never once found the time or place where I was willing to doff my skirts and appear in bloomers. I had a medium-short skirt, properly cut, just as easy to ride in as bloomers, and one certainly feels more comfortable in such as costume, when off one's wheel. I was always treated with kindness and courtesy, and I attribute a great deal of it to my skirts.
>
> Le Long 1898b, 596

In an account entitled "A Honeymoon on Wheels," a fictional bicycle honeymoon through the South, the bride was told in New Orleans "I reckon a girl in bloomers would ride but once here. She'd get lynched!" (Follett 1896, 5–6).[8] In Chicago, creative entrepreneurs found a market for public dances where "cycling costume was required for both ladies and gentlemen" but only one dance was held before Chicago police passed a ruling: "any woman who would wear bloomers while dancing in public was to be treated as a common prostitute" (Smith 1972, 103–4).[9] A suitable costume for cycling was a significant issue for many women cyclists. The Pope Manufacturing

Company, makers of Columbia bicycles, created paper dolls so women could visualize different cycling costumes (Macy 2011a, 53). What emerged from the debates over cycling dress was not a singular cycling costume but great diversity: each woman found an outfit to wear cycling that was functional, comfortable, and even fashionable.

It is difficult to put an exact number on the adoption of bicycling by women, but it was quickly adopted by a broad spectrum of Americans. Gary Tobin examined League membership applicants and found they were largely "clerks, bookkeepers, businessmen, and professions" (Tobin 1974, 840). Working girls, too, took to bicycling and for far more than just leisure:

> Last winter, during the street car strike in Philadelphia, woman used their wheels as a means of locomotion, and the crowded down town streets were filled with them... . Clerks in stores, typewriters, and the whole great army of employed women rode their wheels to business; women who came to buy left bicycles in the check rooms of the great shops.
>
> The World Awheel 1896, 158–9

Writing in 1952, Sidney Aronson estimated that the total number of bicycles in use in the 1890s was "ten million in a population approaching seventy-six million" (Aronson 1952, 309). In 1899 alone, the U.S. Census reported that "more than 1.1 million bicycles, with an aggregate value of more than $33 million, were produced by more than 300 firms" (Petty 1995, 32). A boom in bicycle manufacturing led to a significant drop in prices; between new and used bicycles, bicycles were within reach of most Americans.[10] Writing about the decline in bicycling in 1901, "H.T.P" comments: "Three or four years ago a good model cost a hundred dollars, but at the present time it can be purchased for a quarter of that sum, while a very fair one can be had for even less" (H.T.P 1901, 426).

But it was not just the bicycle; the bicycle was at the forefront of a consumption boom that involved not just the object but everything that went with it. Bicycles were advertised extensively in magazines and newspapers. Soon the bicycle ads were joined by advertisements for tires, lamps for night riding, seats, bells, toolkits, cycling costumes from head to toe, especially for women, skirt guards, bicyclists' firearms and more. Americans became cycle-crazy with plays, songs, and popular novels and short stories written about the bicycle.[11] The 1890s are exemplified by the classic song "Daisy" where the young man asks Daisy to join him: "it won't be a stylish marriage, I can't afford a carriage/ but you'll look sweet upon the seat of a bicycle built for two."[12] The *New Orleans Picayune* reported in 1895 that Americans spent fifty million dollars annually on bicycles and equipment (Rogers 1972, 203). Bicycles and their related sundries were extensively advertised in magazines, at the start of mass advertising, with a single issue of a magazine, such as *McClure's* or *Scribner's* having as much as 40 pages of advertising to bicycles and related materials (Petty 1995, 34).

Needing information, sharing information

There was a wide range of published information on women's cycling. There were journals published about cycling, such as *Cycle Age* (for "Wheelmen and Women") and *Wheelwoman*. *Outing* magazine focused on all forms of recreational activities, from hunting and fishing to cycling and eventually automobiling. Pope Manufacturing published a brochure in 1892 on "Cycling for Ladies." Magazines produced special bicycle issues, such as *Godey's Magazine* (1896) and *Munsey's* (1896). Manuals on cycling for women were published, such as Maria Ward's *Bicycling for Ladies* (1896), Lillias Campbell Davidson's *The Handbook for Lady Cyclists* (1896) and Miss F.J. Erskine's *Lady Cycling: What to Wear & How to Ride* (1897). And books and novels about cycling were widely available, such as Frances Willard's *A Wheel Within a Wheel* (1895/1991) and novels such as the low-brow comedy *Betsey Jane on Wheels; A Tale of the Bicycle Craze* (1895) and H.G. Wells' romantic comedy *Wheels of Chance: A Holiday Adventure* (1896). Through such texts, women were exposed to the pleasures of mobility.

Information on cycling was circulated to the American public, advocating for cycling and positioning cycling as appropriate for women.[13] As is made clear in Frances Willard's text, the bicycle "… was not only a domesticator, it *was* domestic, a parlour on wheels. It had its own code of conduct, manner of dress and decoration" (Mackintosh and Norcliffe 2007, 172).[14] Through bicycle advertisements, women are depicted as elegantly yet modestly dressed, "… providing constant visual reassurance that women could ride the bicycle with grace and even modesty" (Garvey 1995, 70). As women ventured out, there were societal concerns that greater freedom and mobility might result in women losing their feminine traits, such as "'delicacy,' 'tenderness,' and 'virtue'" (Strange and Brown 2002, 616). Thus, cycling manual such as Maria Ward's, books such as Frances Willard's, articles in *Outing* and *The American Wheelmen*, all emphasized "domestic cycling," a practice of cycling that compromised neither femininity nor family life. While bicycling had facilitated the transformation of American women's lives in terms of their dress and mobility, it was still very much couched in terms of conservative notions of femininity (Finison 2014, 63).

Once what to wear and how to ride had been established, the next question was where to go:

> But when my lady has forgotten all these trouble-some tricks, when she can mount quickly and expertly and do her ten miles unweariedly, where shall she ride? Ah, where? The smiling countryside holds out arms of welcome to her, the shaded grassy road, the smooth steep incline, the bumping corduroy by-ways, the canal towpaths, the lakeside drives and the stubborn still hill to be climbed.
>
> Denison 1891, 54

In bicycling articles, women are given outright suggestions on where to ride. Early adopters of bicycling in major cities rode in winter on indoor tracks created for cycling (Spreng 1995; Townsend 1895, 706; The World Awheel 1896, 155).[15] In nice weather, women cycled in safe public spaces, such as city parks and certain boulevards. As technology improved, making riding more comfortable and safer, riders began to venture further. Newspapers and magazines provided directions for bicycle rides within range of the city (Tobin 1974, 842). Marguerite Merington in "Women and the Bicycle" not only provides suggestions on what women should wear but also where to cycle in New York City: "From Christmas to Christmas Central Park is a favorite haunt of the cyclist when the weather is kind… . Riverside Drive and the Boulevard offer fair roads and a breeze coming fresh from the sources of the Hudson, untainted as it sweeps by Albany: the historic ground of Washington Heights is practical as well as picturesque… ." (Merington 1895, 704). Given that most bicyclists were middle and upper class and in urban areas, cycling was largely promoted in cities and in adjacent rural areas. The poor quality of rural roads made bicycling for farmers and their families a challenging proposition.

Visual cues on where to cycle were given in advertisements and in illustrations. Many of the bicycle advertisements are set in natural settings, with trees and landscapes visible behind them (Figure 2.2). Frances Willard's *A Wheel Within a Wheel* is illustrated with photographs of her learning to ride, all are garden settings or rural (Willard 1895/1991). Seldom are women depicted as bicycling in cities, with the exception of parks. One of the frequent arguments for women cycling was that it would permit a return to nature and the outdoors, promoting good physical and mental health. Through advertisements, bicyclists were imagining themselves on cycles and moving through these landscapes (Bogardus 1998, 517). It was also visually offering suggestions on where women should ride.

Another source of information about where to ride was the League of American Wheelmen. The League was established in 1880 to promote the rights and interests of cyclists. Instrumental in its founding was Albert Pope, "founder of the U.S. bicycle industry and the Columbia brand" (Petty 1995, 39). Pope was active in establishing local clubs and the first national organization of bicyclists, the League of American Wheelmen. He published books and magazines, such as *Wheelman* (which eventually merged with *Outing* magazine). He lobbied for better roads, with his cycling magazine merging with *Good Roads* magazine. Pope even offered prizes to physicians who published articles on the positive health benefits of cycling. Pope encouraged women to cycle but the position of the League on women members is muddy. It appears that individual League branches would admit women members, such as the Winnebago Wheelmen (Fond du Lac, WI) who voted and passed a resolution in 1896 admitting women members with twenty-eight women then joining (Julka 1941). But the League did have a firm prohibition against women's racing and against members that were anything but "white wheelmen" (Finison 2014; Ritchie 2003).[16] By 1898, the League had over 100,000 members (Strange and Brown 2002, 612).

The League began to map out the roads and their conditions and provide maps based on member's log books to their members (Figure 2.1). Non-members could purchase the maps but the price encouraged people to become members. In 1897, membership fees were $2.00 the first year and then $1.00 each succeeding year, for which you received, in addition to membership, a subscription to the journal, a state tourbook, price breaks on hotels, restaurants, and repairs, and a complete set of road maps (*Wisconsin Tour and Road Book* 1897, 115–7).[17] The price of the maps alone to non-members was $2.00. Other publishers produced bicycle maps but followed the League's example and keyed the roads on the maps by their conditions and grades. Skilled map readers could then "see" the "lay of the land" as they read the map. They could plan their route according to where they wished to go, the road conditions they were willing to endure, how steep the roads were, and other details. If they incorporated the tourbook into their planning, they could add restaurants or hotels that were League approved. As bicycling boomed, so did the demand for spatial information:

> Go where you would you could hear nothing but talk about different models and 'century runs' and the condition of the highways; while map-makers reaped a small fortune by publishing little guides and roadbooks for the use of the bicycle fiend.
>
> H.T.P. 1901, 425

Bicycle touring accounts began to appear in a variety of forums, either in the form of travelogues or in short stories. While some travel accounts were written by men and others by anonymous contributors, attributed to "Martha" or "the Prowler," significant numbers were written by women for women. Through written accounts of women's cycling adventures, readers were "modeled" not only how to ride but how to go about planning a bicycle trip. In a column on "Cycle Touring" from June 1898, directions are provided on planning a cycling trip:

> To lay out a cycle tour, first roughly sketch out a route from the map. Then, from the road-book and map combined, make entries in a conveniently sized notebook of all the towns and villages to be passed through or by, with the distance of each from the starting point, and in a parallel column the distance between each intermediate place and the next one. After each entry leave sufficient space for remarks.
>
> "The Prowler" 1898, 320

Restaurant and hotel information made it clear that food and lodging were available all along the route. Some hotels and restaurants even offered a discount for League members (Tobin 1974, 843–4). It was essentially taking the uncertainty out of travel, so that the travelers could focus on the pleasures of the trip.

Women were encouraged to take on longer trips of a week or more bicycling through parts of the United States or Europe. Magazines like *Outing* and *Good Roads* were saturated with articles that described tours that could be done in a weekend, more substantial tours throughout the United States to places like Yellowstone or the Rocky Mountains, and even more ambitious tours of Europe (Tobin 1974, 842). These tour descriptions could be read as how-to guides or for pleasure. They often involved either a trip by a group of young women or a married couple and described how they went about their tour, from the planning to the actual trip itself. Guide books indicated not only routes and road conditions: they also included information designed to make longer trips as pleasant and successful as possible. Women cyclists were directed to seek out these guide books:

> Road-books with best routes and conditions of the highways are issued to all members. These contain, also, a list of the best hotels to patronize (often at special rates to cyclists); lists of repair shops, the names of 'consuls,' who are appointed in every chief town and who willingly impart to strangers all necessary information.
>
> Dodge 1888, 210

Railway information was provided to get cyclists to the start or back from a tour. Restaurant, hotel information, bicycle shops/repair shops were all listed in the guidebooks. Most guidebooks also had advertisements in the front and back of the guides. While the guides clearly state that listed businesses are vetted by League members, there is a fine line between the advertisements and the "vetted" businesses.

American Grace Denison published an *Outing* account of cycling in Ireland in 1893 (Denison 1893a–e). She was inspired to take the trip from a newspaper account of an Irish cycling club. Denison writes "the more I thought the more I fretted to set out. As to the roads, I cheerfully re-read the newspaper article, and in fancy followed the '*Ohne Hasts*' on their runs (Denison 1893a, 28). She goes on:

> After the costume and the wheel, came the itinerary. (Isn't that sequence a confession of sex?) I had maps and advice galore, a good share of undiluted Hiberian pessimism, with large portions of much watered veracity, and highly colored adjectives.
>
> Denison 1893a, 30

Not only is she modeling how to plan and implement a bicycle tour, she is standing in for the reader with her introduction where she describes reading an article about cycling in Ireland and begins to plan her own trip. She concludes her five-part travelogue with a call to her readers: "Nothing remains but to say to my sister wheels, 'Go and do likewise'" (Denison 1893e, 447).

"Martha" published an account in 1892, "We girls awheel through Germany," of a trip through Germany by five women on bicycles, describing the landscape and the people but also explaining how they found inns and food, dealt with bicycle repairs themselves, and what they packed in their bags: "changes of linen, handkerchiefs, necessary brushes, combs, etc., a pocket dictionary, a road map for cyclists, and a few other things" ("Martha" 1892, 302). It concludes: "If you want to see the country and the people in the best possible ways, and if you want to have a delightful time, buy a bicycle and tour" ("Martha" 1892, 302). Through accounts such as "Martha's," women cyclists were shown not only that women could tour along on bicycles but how to tour – what to pack, what to see, what to do, how to do it. The bicycle literature constructs the association of bicycle maps with "touring" in Europe. By modeling the behavior, it was demonstrated that "touring" was within the reach of the middle-class in the local landscape if they had a bicycle and a map.

Besides directions on how to travel, these accounts also provide a travelogue of sorts, providing a sense of the landscapes that the riders are experiencing. For example, from Grace Denison's "Through Erin Awheel":

> We got on to Powerscourt in peace and satisfaction, and on producing an order, which we had got for the asking from Lord Powerscourt's agent, we were admitted by an ancient lodge-keeper into the beautiful demesne, and took a ride through that sixteen thousand acres which own Lord Powerscourt as lord of the soil. Of course it goes without saying that the roads here were something perfect, and on either side clumps of rhododendrons, white, pink, and crimson in full bloom, ferns and wild flowers, rustic bridges and shaven turf, grand old trees, and many a tinkling rill, at the sound of whose silvery voice we always fancied we heard the first echo of the far-famed waterfall. But sixteen thousand acres gives quite a lengthy ride, and we surfeit with glimpses of lovely places, soft lawns, bosky dells, dainty deer flitting under the great oaks, or staring knee-deep in the feathery ferns, before we came to the foot of the cliff and in sight of the little lace-like fall, leaping and dashing from a rock seventy feet above us. This is one of the picture places of Ireland... .
>
> Denison 1893a, 32

Or from Mrs. Alice Lee Moqué's "A bohemian couple wheeling thro' Western England":

> How can a few necessarily brief sentences do anything like justice to dear, quaint old Chester, its cathedral, its Roman walls and gates, dating back nigh two thousand years, and its many ancient streets crowded with historic interest?
>
> Moqué 1897, 190

Alice assists her readers in visualizing the landscape by comparing it to more familiar lands: "The scenery in many parts of England puts one in mind of the rolling districts of Maryland and Virginia. The fields are well worked—unlike many of ours—and the flocks grazing on the green, grassy slopes and numberless cows, with tinkling bells, make a picturesque memory" (Moqué 1897, 190).[18] It is remarkable that in the accounts that I have read, there are dubious lodgings, poor food, bicycle problems (flat tires, broken chain guards) and inevitable close calls (runaway bicycles, near crashes with farm animals, etc.), but I have yet to see an account where a woman describes being lost.

Gender and mobility

[T]he average bicycler has learned more of the topography of the country and its local history than anybody, except the map makers, has known before. The feminine mind is fond of detail, and when she starts out, the woman who rides afield knows where she is going, and what she is going to see. It is as though a new language had been given to her, and the books of literature opened before her.

The World Awheel 1896, 159

In the 1890s, there were definite gendered expectations when it came to mobility. Culturally, men were associated with the public sphere and expected to move through the world. Women were associated with the private sphere and with home and family. Marguerite Merington, writing in 1895 stated: "Now and again a complaint arises as to the narrowness of women's sphere. For such disorder of the soul the sufferer can do no better than to flatten her sphere to a circle, mount it, and take to the road" (Merington 1895, 703). Men adopted the bicycle first, given the challenges of the Ordinary bicycle, and given their gender roles at the time, and were associated with bicycling greater distances (such as "century runs" or 100 miles in a day) at faster rates. The media associated male cyclists with action, adventure, aggression. Female cyclists were told by society through the media to be lady-like, proper, cautious, stately, modest. Both men and women were told in magazines and cycling guides: Don't be a scorcher! That is, don't ride fast and aggressively and be a threat to others. For example:

The feminine scorcher is not an altogether lovely object; unless she intends to be a professional, she should leave this form of amusement to men. A woman with her back doubled into a bow-knot, her hat awry, her hair disheveled, and her fact scarlet with exertion, is neither fascinating nor attractive; she takes on an anxious, worried look in her eyes, has her muscles developed at the expense of her feminine grace, and her complexion coarsened by the rude contact of wind and weather.

The Feminine Scorcher 1896

While women were warned against being "scorchers," there was male fascination with them, resulting in poems and songs about the subject, such as the popular song "The Pretty Little Scorcher." Male scorchers were somewhat more acceptable than women scorchers, as long as they did not threaten public safety. And as for the century rides: "The Century Ride is scarcely to be undertaken by women in general; fifty or seventy-five miles a day is as much as the average woman can endure" (Wheel Whirls 1896). How bicyclists moved through the landscape was culturally gendered with appropriate styles for men and women.

Guides and tour articles were initially written by and for a masculine audience, given American men's early adoption of cycling. Some cycling directions offer asides to women cyclists, such as in the "Cycling" column in *Outing*, where in the discussion of a tour in the Northern Atlantic states includes a paragraph addressed "to the lady tourist" comments on how to handle luggage, where to pack your camera, and encouragements about the rewards of bicycle touring ("The Prowler" 1896, 43). Similarly, on the 1896 "Cyclists' Road Map of Portland Oregon," in addition to information about roads, road quality, towns, taverns, sense of topography, there is a note in the advertising border reading:

> For ladies and those not wishing to make the round trip by wheel, a delightful day may be had by taking the O.R. & N. steamers from Ash Street at 7 A.M., arriving at St. Helens about 9. You then have nearly the whole day in which to enjoy the beautiful scenery along the road, and the lunch you have taken may be daintily served at one of the numerous sparking streams to be found about eleven or twelve o'clock. This run should be made when the wind is from the north. Fare to St. Helens, 50 cents, wheels free.

Both *Outing* magazine and the map are writing to a diverse audience, assuming that women would be reading them as well as men. This is in addition to the articles and stories written to appeal to a feminine audience, such as Grace Denison's "Through Erin Awheel" or "Martha's" "We Girls Awheel through Germany" (Denison 1893a–e; "Martha" 1892).

While the media constructed an image of dainty women touring, evidence from cycling clubs and from accounts such as LeLong's tour, provides glimpses of women actively engaged in cycling. A "Souvenir Programme of the Bicycle Runs of North Side L.A.W. Club, July 5th to 10th, 1897, Milwaukee, Wis." details the bicycle runs and the guides. Twenty different runs were offered of which two were listed as "Ladies' Runs." "Ladies' Runs" suggests a gendered difference in the runs but no information is provided to clarify the difference between a regular run and a "Ladies' run." However, of the ten cycling guides listed in the program leading the runs, six are quite definitely women, identified by "Miss" in front of their names. As there were only two "Ladies' Runs," it seems reasonable to assume that ladies were not restricted to "Ladies Runs," they were leading the other runs too that crossed the Milwaukee landscape.

An account of cycling in New Jersey suggests that women may have the upper hand when it comes to cycling and geographic knowledge: "Indeed, past experience has taught the masculine section of our touring club to allow the gentler sex a free rein in the matter of choosing routes, for invariably their instinct, reason, or whatever faculty they bring into service when dealing apparently hap-hazard with a geographical problem, prompts them to decide satisfactorily" (Godfrey 1898, 8). The quote at the beginning of this section goes further to suggest that women have a natural affinity for geographic knowledge and that with access to landscape and opportunities: "It is as though a new language has been given to her, and the books of literature opened before her" (The World Awheel 1896, 159).

Mobility and geographic information

Early forms of geography were in the form of descriptions of places, especially distant locations. Many of these accounts were written by geographers who had traveled widely but who also often utilized the travel writings of others, such as the works of the Greek geographer Strabo or the Muslim geographer Ibn Battutah (see Cresswell 2013, 14–34). These descriptions became the basis of geographic knowledge, often including maps or the directions for creating maps. As geography became a formal subject of study, mobility continued to play a role as geographers sought to understand the commonalities and differences between locations, to facilitate movement between places (trade, ideas, people), and to comprehend factors influencing connectivity and accessibility.

The travel accounts produced by writers such as Merington, Denison and Workman advanced the geographic understandings of their readers in two ways: through promoting women's mobility and through cultivating their geographic imaginations. In promoting women's mobility, they were modeling how women could pursue similar trips, describing how to go about it and helping them to envision themselves doing the same. By the 1890s, American women such as Elizabeth Cady Stanton and Susan B. Anthony had been working for over forty years to try to improve the lives of American women but had only limited success. The bicycle brought about two significant changes – mobility and dress – that had been hoped for but previously not achieved.

As women increased their mobility through bicycling, not all welcomed them. If the woman cyclist had the audacity to wear bloomers, they were often subject to unwelcome attention, such as in 1895 Fond du Lac, Wisconsin:

> A Chicago lady attracted considerable attention Tuesday evening, by appearing at the fire near the Central depot in light tanned bloomers. For a time it was impossible for anyone to locate the fire as a crowd had

congregated at both ends of the platform, some bent on seeing the fire others on watching the young woman, who is Miss Dell Smith now of Chicago, formerly of this city. The lady is a good rider and made a fine appearance.

Julka 1941

One young lady in Chicago in 1895 stopped traffic in her costume of "a flesh-colored sweater and a pair of black tights that left few contours of her form to the imagination. She was promptly arrested on a charge of disorderly conduct and the police magistrate who inspected her costume fined her twenty-five dollars" (Smith 1972, 105). But it was not just their dress, it was the mobility itself that was threatening: any woman who moved too freely bucked gender conventions (Scharff 1999, 294). "The Women's Rescue League," crusaded against women bicycling arguing: "Bicycling by young women has helped more than any other medium to swell the ranks of reckless girls, who finally drift into the army of outcast women of the United States" (Crusade 1896). As time went by and more and more women took to cycling, it became less and less a novelty but the new norm. As prices of bicycles dropped, more and more women could afford it, and cycling shifted from a plaything of the wealthy to a craze of the masses.

The mass adoption of the bicycle did seem to reach across nearly all economic and social classes in the United States, but this is not to suggest that there were not exclusions. Race was certainly an issue. Black Americans of the 1890s did bicycle but there were spaces they were excluded from, such as the League and League events (Finison 2014). And, as Virginia Scharff writes: "We should not mistake female mobility for emancipation" (Scharff 2003, 6). Women had greater mobility but they still had (and have) a ways to go before any claims of emancipation can be made.

Further, as women went out and had their own cycling adventures, they created their own geographies, constructing their own worlds through their experiences (Watts 2008, 719). Their geographical knowledge was expanded through the experience of cycling. It is not just a matter of going from point A to point B. Bicycle trips were often discontinuous despite the rider's best intentions. Once the journey begins, there may be stops for environmental conditions (rain, wind, mud), technical difficulties (flat tires, skirts caught in chain or wheels), stopping for trains to pass or other traffic. And there is the haptic experiences of riding in the elements – of hot sun, of the cool shade of entering a treed area, of bumpy roads or of a very smooth road, of burning muscles as a hill is climbed. The practice of bicycling involves the consumption of place – for women this meant venturing out into the landscape in a new way. As women cycled, they came to know their local roads: "Gradually I began to get intimate with the roads and to know all their little ups and downs, like wrinkles or like a family disposition" (Rudd 1895, 126). This knowledge was intimate, of the little bumps in the road, of the familiar sights and smells along their route, of where to stop for a cool drink of well water. Through these experiences, riders develop route-based wayfinding knowledge, composed of the sights and sounds of traveling along a particular route, and further refined by repeat trips (Golledge 1999, 7–9).

Cycling and the American landscape

The cycling triggered a transformation of the American landscape, with cycling actively producing spaces. Cyclist organizations, led by the League of American Wheelmen (LAW), were among the first to critique the American infrastructure of roads and highways and call for improvement. The League began calling for road improvements early. Prominent New York member Isaac Potter published *The Gospel of Good Roads* in 1891 (he would eventually be elected president of LAW). In it, Potter argues that the improvement of roads was in the best economic interest of both merchants and farmers and that it was the responsibility of "the State" to make and repair roads. This was just the start. In 1893, Congress was pressed by LAW members to appropriate $10,000 to study road improvements. The League began publication of *Good Roads Magazine* in 1895. Potter published articles in prominent magazines, promoting the Good Roads movement, such as in *Harper's Weekly* (1896) and *Godey's Magazine* (1896). It was the Good Roads campaign that "quite literally paved the way for the automobile" (Rogers 1972, 201). States began appropriating money for roads and establishing highway commissions: "In Massachusetts, all the members of the State Highway Commission belonged to the L.A.W." (Aronson 1952, 310). The League also constructed signposts with travel information, beginning the process of marking the roads and providing basic directions (such as number of miles to a particular town). League maps were created from members' log books (Akerman 2002, 178; *Road Book of Massachusetts* 1898; *Wisconsin Tour and Hand Book* 1897). Some regional and state branches of the League did admit women members, so it is conceivable that women contributed to the creation of LAW maps but essentially impossible to prove. It is clear though, that through articles such are Margaret Le Long's tour from Chicago to San Francisco, greater attention was brought to the condition of state roads and the need for improvements, as well as for the need for better maps of these roads. In addition to improved roads, bicycle touring began the process of creating "corridors of consumption," which boomed later with the automobile: "Wherever the tourists went, a service corridor was built to serve them. Incorporated into these travel corridors were the urban amenities for physical comfort and social segregation plus abundant open space. The tourist could see flowers, stop and reflect on the green grass, breathe fresh air and still have sociable company and comfortable lodging" (Tobin 1974, 845). In areas such as New England, in close proximity to major cities such as New York and Boston, cycle tourism resulted in the expansion of businesses catering to cyclists brought in by the promotion of the region through cycling literature. Inns, restaurants, bicycle repair shops, and railroad lines advertised in cycling magazines and League guide books.

"Free untrammeled womanhood"

Through this greater mobility, American women were consuming new forms of geographic knowledge. Some of this knowledge was practical while for some readers, this geographic knowledge was "virtual." Relatively few who read the travel accounts would actually go and tour in New England, or Germany, or India. These descriptions of travel, with their accounts of peoples and places traceable on maps suggest the development of a geographical imagination. Through their mental maps and imaginations, they would follow the routes, whether they actually toured or not. A British poem entitled "The Old Road Map" captures how a map was used not only to plan a tour, but became the simulacrum for a trip once taken:

> That summer is dead, and its friends are fled: but I
>   follow their passage still,
> Where the pencil-black still wanders back from Wells
>   to Cranmore Hill:
> It was here that we lay after lunch that day for a pipe
>   and a noontide nap,
> Till the market-cart woke us all with a start, as we
>   dozed o'er the old Road Map.

Waugh 1900

By 1900, the bicycle fad was fading. Prices of new and used machines dropped (Petty 1997, 125). While it is unclear why the bicycle craze ended, some scholars suggest that it may be because of oversaturation of the bicycle market (Pridemore and Hurd 1995, 88–90). Others point to the development of automobiles.

The impact of the bicycle on American women's lives was significant. The bicycle achieved in five years what dress reformers had been seeking for fifty years (Aronson 1952, 308). Both Susan B. Anthony and Elizabeth Cady Stanton, who had been fighting for women's rights for most of their lives, wrote and published essays where they commented on the advancements they saw in women's rights largely tied to the bicycle. Stanton, writing in *The American Wheelman* in 1895, call it "one of the greatest blessings in the 19th century" (Stanton 1895). Anthony, interviewed by Nellie Bly in 1896, said:

> I think it has done more to emancipate women than anything else in the world ... away she goes, the picture of free, untrammeled womanhood.

Bly 1896

Girl at the wheel: Women and the automobile

Alice Ramsey, like LeLong, crossed the country in the latest technology – now the automobile – in 1909.[19] Ramsey was the first woman to drive across the United States, via what became the Lincoln Highway (Ramsey 2005, 1). Wayfinding information was essential for the entire trip. East of the Missouri River, *Blue Books* were available for eastern states but leaving, according to Ramsey "a vast void in the great wide West" (Ramsey 2005, 240). *Blue Books*, beginning publication in 1901, were the most up-to-date travel directions of the time, providing written directions from one point to another within states (Bauer 2009). But Ramsey found them at times wanting, like LeLong did with the LAW maps. And like LeLong, Ramsey encountered poor roads all across the United States with Iowa particular appalling, compounded by unseasonable heavy rains. Interestingly, Ramsey lost the road in Wyoming a number of times too (Ramsey 2005, 91–2 and 102).[20]

Unlike Le Long, Alice did not travel alone. While Alice drove the Maxwell, filled with her "crew" of her husband's two sisters (both in their forties) and a young friend of Alice's (who, like Alice, was in her early twenties), they were accompanied by a pilot car and had an "advance man" facilitating the trip (McConnell 2000, 23). Of the four women, only Ramsey knew how to drive, so she did indeed drive the whole way. Ramsey was the veteran of a variety of automobile races and runs and was challenged by her local Maxwell automobile agent, who saw a promotional opportunity. Automobile manufacturers sponsored endurance runs, road races, and cross-country tours to actively encourage customers to consider long-distance travel (Akerman 1993, 11).

As the cross-country trip was not a race, Alice and her crew could take their time. They were wined and dined across the country and treated as celebrities, which in a way they were. Their trip was well publicized by the Maxwell-Briscoe Company, who provided their touring car and funded the effort, and newspapers all across the country followed their itinerary. It would take them two months to go from New York City to San Francisco (June 9th to Aug. 7th).

Alice was the first of a series of women to drive across the country. According to Curt McConnell there were four benefits of women's cross-country automobile runs: demonstrated the utility of the automobile and promoted it; documented poor road conditions and the need for improvements; established that women were competent drivers; and revealed to manufacturers that women were part of the automobile's consumer audience (McConnell 2000, 1). Such tours were the most visible embodiment of women's increased mobility.

With the invention of the automobile, women's mobility expanded exponentially over the next forty years. Initially, the automobile was limited to the wealthy Americans who had the time and finances to engage in this new "sport," as it was initially seen. By the end of the Progressive Era, the automobile was within reach of many Americans, and no longer seen as just "sport." This greater spatial mobility made possible by the automobile resulted in the need for and the circulation of greater spatial knowledge, a process that was initiated by the bicycle but multiplied exponentially by the automobile. It transformed the American landscape profoundly, with women playing an active role.

Women and automobility

> Motoring and, indeed, all outdoor pleasures are masters of iron, especially to a woman, who always, in a fashion, has to adopt them, while a man takes to them naturally.
>
> Why women are ... 1904, 156

The bicycle industry in the 1890s was the "training ground" for some of the earliest automobile makers in the United States with many manufacturers beginning with bicycles and advancing to automobiles as they developed new technologies. Albert Pope began with manufacturing a variety of small articles, such as an air pistol and cigarette rolling machine before mastering bicycle manufacturing in the form of the Columbia (Epperson 2000, 303).[21] Pope Manufacturing began work on a "Columbia horseless carriage," but made a "strategic miscalculation" and "failed to adapt," eventually folding in 1913 (Epperson 2000, 312 and 316). Charles and Frank Duryea began with bicycles in Chicopee, Massachusetts, read about the new European horseless carriages, and produced their own "motor buggy" in 1893 (Scharff 1991, 8–9). A more advanced Duryea model would win the nation's first automobile race, a round-trip between Chicago and Evanston, Illinois in 1895.[22] The first automobile to be manufactured in quantity was an Oldsmobile, crafted by Ransom E. Olds, with 600 produced in 1901 and, by 1904, over 5,000 vehicles (Scharff 1991, 11). Early automobiles were expensive, costing over $1000, a sizeable sum at the time and beyond the reach of most Americans. As a result, they became "a symbol of conspicuous consumption" and toys of the wealthy, who often did not drive the automobiles themselves but used chauffeurs, often former coachmen (Scharff 1991, 17). Those men and women who took on driving had to learn not only to drive but also learn to keep the engine lubricated, repair tires, trouble-shoot engines, if they were to be an "automobilist" (Clarsen 2008, 13). An 1898 advertisement for the Winston Motor Carriage proclaimed that it was "Better than Horse or Bicycle. One of the most delightful of modern possessions... ."[23] By 1900, there were 8,000 registered automobiles in the United States.

But registration figures do not reflect the actual number of drivers in the country. Not all states required driver's licenses or the registering of automobiles at this point, so it is difficult to pin-down exactly how many drivers there were by 1900. But women were among the owners and drivers. Julie Wosk writes that "In the United States, Genevera Dephine Mudge, of New York City, and Daisy Post were driving in 1898, and Mrs. John Howell Phillips, of Chicago, was reportedly the first American woman to obtain a driver's license, in 1899. By 1902 thirty-five Chicago women, most of whom were young and half of whom were single, had obtained licenses" (Wosk 2001, 116). In 1900, Anne French had to qualify for a steam engineer's license before driving her Locomobile in Washington, D.C. (Clarsen 2008, 13). Alice Ramsey was given an automobile by her husband in 1908 after a runaway carriage incident with his powerful bay horse. He never did learn to drive while she quickly put 6000 miles on her car (McConnell 2000, 25). A 1900 *Outing* article on "The Place of the Automobile" states that "A very large proportion of the men and women now enrolled in the ranks of the automobilists owe their intimacy with the new sport to personal and persistent liking for just that sort of ambition" (likening to the ambition to drive a locomotive engine) (Bruce 1900, 65). The use of "men and women" at this early date, 1900, suggests that it was not just a man's pursuit. The article features six photographs, two of which are women at the throttles of their electric automobiles. Electric automobiles were marketed especially to women, framed as quieter and cleaner than gasoline models while not having the power and range (Clarsen 2008, 14). Gasoline automobiles were seen as "powerful, complicated, fast, dirty, and capable of long-distance runs" and "belonged to men" (Scharff 1991, 35 and 37). Women early adopters of the automobile were more often driving electric. But this was soon changed, for women quickly turned to gasoline cars, especially with the development of electric starters in 1912, with electric automobiles falling out of favor by 1920.[24]

The author of "The Place of the Automobile," Robert Bruce, attempted to put into words the appeal of the automobile:

> Above all else in point of real enjoyment is the feeling of being lifted up and along – carried away – rather than that of being drawn or pushed. The absence of the horse from in front emphasizes the difference for a while; then it becomes a mere matter of course. The mechanical capacity to negotiate from fifteen to thirty-five miles an hour on favorable stretches is the smallest part of it. Mere distance-covering has long since been dismissed from the list of recreations.
>
> Bruce 1900, 66

Bruce goes on to discuss different models and the differences between American and European motorists. He does point out much depends on road conditions and that if they are improved "then we may witness a return to them of undreamed magnificence, and our people of wealth and leisure may move from country seat to country seat and tour from city to city, or from one social resort to another in all the pomp and circumstance that lavish expenditure can buy and emulation spur them on to" (Bruce 1900, 69). Bruce clearly is viewing automobiling as a recreation of the wealthy, "a new sport," but in his last paragraph he does recognize that "there is no apparent limit to the use, or usefulness, of the automobile" (Bruce 1900, 69).

A humorous *Outing* essay on "Why women are, or are not, good chauffeuses" from 1904 addresses the desire of American women for automobiles. "The Little Woman" longs for a gasoline motor, but men at the club question whether she is up to it. A woman driver comments "Has one to be shorn of all feminine graces, then, to be a good chauffeuse?" The answer in the essay appears to be "no, but ..." with the view expressed that men take to them 'naturally' and women have to "adopt" them (see quote at the start of this section). Impulsiveness, nervousness, lack of concentration are all negative "women's traits" that preclude a woman from being a "good chauffeuse" but "If she can play the piano, if she can play bridge, if she can paint, even if she can typewrite, she will, ten to one, be a good chauffeuse" (Why Women Are ... 1904, 157). The intent, obviously, was to have women read this essay and think: I can do all of these things, I can motor.[25]

While I have included a number of articles from *Outing*, a magazine focused on sporting and recreational activities, many other magazines covered the automobile. Women's magazines, such as *Good Housekeeping* and *Ladies' Home Journal*, featured articles and stories tied to automobiling, such as "The Girl Who Drives a Car," "Little Things About a Car: That Every Woman Who Means to Drive One Ought to Know," and "Woman at the Motor Wheel" (Hitchcock 1913; Little Things ... 1917; Murdock 1915). Articles such as these emphasize being trained by a professional, learning to evaluate your car and fluid levels before every drive, becoming attuned to the engine, and, above all, the importance of "patience and perseverance" (Hitchcock 1913, viii).

In addition to articles, advertisements also promoted women as motorists (Figure 2.3). Ads targeting women through text and images and through their placement in women's magazines. Early automobile advertisements targeting women likened their operation to that of home devices – "As simple to run as a sewing machine" or "As easy to operate as a kitchen range" (Clarke 2003, 110). Such analogies not only reassured women that they could drive but that the automobile was, in a way, a domestic technology. An Oldsmobile ad from 1905 states " Makes everyone your neighbor" and includes in the ad a range of models and prices, from the Standard Runabout, $650" to the "Oldsmobile Heavy Delivery Car, $2000."[26] Waverly was running advertisements for electric vehicles as early as 1902 in the *Ladies' Home Journal*.[27] A small, text-only advertisement for "Automobile Bargains," both new and used, appeared in several 1907 *Ladies' Home Journal* issues.[28] By 1910, the most commonly advertised product in magazines was the automobile (Gudis 2010, 372).

Through texts and images, women were being sold on the automobile, with many benefits extolled, including reducing feelings of isolation. Christine Frederick wrote in 1912 of life as a "commuter's wife" and that living in the country is splendid with an automobile:

> [L]earning to handle the car has wrought my emancipation, my freedom. I am no longer a country-bound farmer's wife; I am no longer dependent on tricksome trains, slow-buggies, the 'old mare,' or the almanac. The auto is the link which binds the metropolis to my pastoral existence; which brings me into frequent touch with the entertainment and life of my neighboring small towns—with the joys of bargains, a library and soda-water.
>
> Frederick 1912

Figure 2.3 This advertisement for the National Motor Vehicle Company of Indianapolis dates from 1916. The model name National "Highway" plays into
1916 interests in national highways. These lovely ladies and their dogs
appeared in *Motor Age*, 13 April 1916, pg. 51.

Besides reducing isolation, the car is tied to living a "modern" life, including such "luxuries" as libraries and soda fountains. In a 1913 article on de-
pression in *Ladies' Home Journal*, the automobile is listed as one the new
technologies "lessening nerve wreck and insanity among the rural popu-
lace... . The automobile, even though the family hasn't one of its own, increases her social opportunities because it brings good roads which lessen
distances and make communication easier" (Adams 1913, 20).

Through articles and advertising, the automobile was promoted and sales
continued to grow. Henry Ford founded the Ford Motor Company in 1903

and began work on an automobile for the everyday American (eventually the Model T) "in a determine effort to lower the cost and increase the diffusion rate of the motor car to all classes of the population" (Cowan 1983, 82). It was introduced in 1908, selling almost 6,000 the first year (Scharff 1991, 52). By 1929, there were 23.1 million registered automobiles in the United States (Seiler 2009, 36).

Sharing information

Women publish information for other women drivers, ranging from how-to manuals, magazine articles and columns, to novels, covering not only the how-tos for driving, but also basic mechanics (Clarsen 2008, 12). Similar to the bicycle, women were encouraged to learn how to do basic repairs, essential knowledge when flat tires were a common occurrence and drivers needed to know how to patch tubes. Alice Ramsey describes herself as being "born mechanical" and over the course of her tour, changed tubes and did other repairs and maintenance, while leaving the repairs of broken axels to the professionals (Ramsey 2005, 15). When Nancy Teape and her daughter Vera drove from Denver to Chicago and back, they were proud of their ability to repair their automobile themselves with what they had on hand. Vera recalled "By the time we returned to Denver our little car was pretty well held together with hairpins and safety pins!" (McConnell 2000, 34). Dorothy Levitt's *The Woman and the Car* (1909/2014) demonstrated that a woman driver could be glamorous *and* know how to repair a car. The photo of Dorothy on the book's frontpiece shows her at the wheel in a fur, looking like a film star (most likely a studio shot). But in the book and its illustrations, she makes it clear that furs are not practical driving, that what a woman driver needs is a good coat, scarf/muffler, and cap, along with an "overall" of linen for when you work on your car (Levitt 1909/2014).

Dress was less a concern than with the bicycle. Women could still wear corsets and skirts and drive a car. Yet there was still a physicality to driving, "an active, whole-body engagement," especially given the poor state of roads and open cabs of early automobiles (Clarsen 2008, 14). Minna Irving wrote in 1909:

> Before I had been riding half an hour, I discovered that motoring discourages all the vanities of fashion. Foolishly, I had put on a big flower-decked hat for my initial spin, draped it, more picturesquely than securely with a long veil, and covered my dress with a silk coat, fastening with flimsy loops instead of good old-fashioned buttonholes. As soon as the machine gathered speed, the wind whipped my coat open and tore off my hat, and I finished the ride looking like a demented Valkyrie.
>
> Irving 1909, 233

Alice Ramsey describes their appearance as they arrived to attend the Cobe Cup Race in Chicago:

> Hats, veils, and dusters failed to keep us protected or clean, and by the time we arrived, people were calling out to us "Oh, you kid!" and "Why don't you wash your dirty faces?" We were truly sights! Where our goggles fitted around our eyes, excluding the dust, there were circular patches of skin several shades lighter than the rest of the face. We look as if we belonged to another *race*! (No pun intended!).
>
> Ramsey 2005, 41[29]

Woman motorists soon learned, whether on their own or from writers such as Irving and Leavitt, that they needed to dress practically for "wind, weather, and dusty roads" as well as to deal with mechanical problems (Irving 1909, 234).

There appears to be fewer articles on how to plan automobile tours compared to bicycle touring. A 1906 *Ladies' Home Journal* article entitled "An automobile vacation on $1.60 a day" explains with pictures how a woman planned a two-week vacation using a rented automobile for twenty-five people (yes, twenty-five!) (Humphreys 1906, 27). More often than how-tos, accounts were published of women's auto tours, such as in 1911 an account of a suffrage tour of Illinois was published in *American Magazine*, detailing how five members of the Illinois suffrage movement went on an automobile tour of Illinois in July 1910. Inspired by a street meeting held at the National Suffrage Convention in Washington, DC, the Illinois women organized an "automobile committee," divided the state into districts and sent out groups "to invade each district, to speak for a week in the parks, streets, and public squares of the towns of each" (Todd 1911, 612). Each party was made up of a group of women who would each address a different aspect of suffrage (such as government, taxation, history, worker's rights). The women, however, did not drive themselves, having a chauffeur so they could concentrate on their presentations, giving as many as five speeches a day. They not only used the automobiles for getting around the state, the automobiles also served as their stage: "We spoke usually from the automobile, driving up into some square or stopping on a prominent street corner which had previously been advertised in the local papers… ." (Trout 1920, 147). This proved to be so successful that Catherine McCulloch publishes a "how-to" in *The Woman's Journal*, "the official organ of the National American Woman Suffrage Association" that same month, describing it as "the best possible method of reaching the man in the street" (McCulloch 1910).

The concept of autotours to promote suffrage became a powerful way for the suffrage movement to get their message out and many more autotours were conducted, including two cross-country tours from San Francisco to Washington DC, sponsored by suffrage organizations. In 1915, the more militant suffrage organization, the Congressional Union, sponsored a cross-country trip from San Francisco to Washington DC, with the car driven by two Swedish immigrants, Ingeborg Kinstedt and Maria Kindberg, accompanied by Sara Bard Field and Frances Jolliffe. Mabel Vernon served as their "front man," traveling ahead and preparing receptions, notifying the press, and lining up speaking opportunities for the travelers. Unfortunately, they did not plan ahead practically: "No one packed a map" and as a result, they spent a night lost in Nevada until they found a ranch "where the travelers were restored with hot coffee, a huge fire in the kitchen stove, a ranchhand breakfast, and—best of all—a map" (Fry 1969, 19–20). As they traveled, Sara gave speeches and the women collected signatures on a national suffrage petition from men and women alike. They ultimately presented their petition with an estimated 500,000 signatures to Woodrow Wilson and to Congress at the 1915–1916 opening day in Washington DC (Fry 1969; Mayor aids suffragists 1915; Scharff 1991, 86).

In 1916, Alice Burke and Virginia Richardson went on an "automobile suffrage circuit" of the United States, sponsored by the National American Woman Suffrage Association (Wosk 2001, 126). Or as it was phrased in a *Motor Age* article on their trip: "more than a transcontinental drive, for their tour included the south and north as well as the east and west" (What Some Women … 1916). They traveled all around the states, from April to October, covering over 10,000 miles, in a suffrage yellow Saxon automobile, donated by the company.[30] Their escapades made local and national news as they covered the country, such as their traveling with a kitten named Saxon and being lost for four days in the Arizona desert. Burke and Richardson made regular stops, giving suffrage speeches and allowing suffrage supporters to autograph their automobile. At a stop in Athens Georgia, the local newspaper reported:

> Both are very convincing speakers and thorough gentle and womanly in their manner. Women like these two do not arouse antagonism even among those opposed to suffrage, because there is nothing unduly aggressive in their manner, though both are so deeply soul and body devoted to the cause.
>
> 'Votes for Women' To Be Discussed … 1916

When they arrived back in New York, they were met and escorted "to Suffrage headquarters by women in decorated automobiles" (Suffrage Autoists … 1916). Saxon Motor Cars then began running an advertisement featuring cameo photographs of Burke and Richardson headlined "Two noted suffragists travel 10,000 miles in Saxon Roadster," such as in *Good Housekeeping* magazine in February 1917.

In a very different auto tour, socialite Emily Post traveled from New York to San Francisco in 1915, publishing a three-part account of the tour in *Collier's Magazine* before producing a book on the tour. Emily Price Post was born and raised on the East Coast in a very well-to-do family. She married man with an even better pedigree, eventually had two sons, and then had a less fashionable divorce (Kolbert 2008). Around the time her marriage was ending, Emily took up writing, publishing stories and series in popular magazines, and eventually publishing a number of novels.[31]

Emily began to plot a trip across the continent to the Panama-Pacific Exposition in San Francisco, which she would document for *Collier's*, her trip inspired by advertisements: "They were all so optimistic; they went to my head! 'New York to San Francisco for $38!' 'Go to the Exposition in a Z----car.' 'Travel luxuriously from your own door through unsurpassed scenery, and over the world's greatest highway, to the Pacific shore'... ." (Post 1915a, 11). But the promises of "the world's greatest highway" seemed dubious, for as she attempted to locate maps for the trip, she quickly learned that information was severely lacking. At the Automobile Club looking for maps, a young clerk encouraged Emily to ask people along the way the best route to take, when she expressed her concern over the quality of the maps.[32]

Ironically, where bicyclist Alice Lee Moqué was comparing the English landscape to parts of the American landscape for her readers, Emily did the opposite, comparing the roads of Ohio to those of France, but not in an entirely favorable way: "they have wonderful foundations, but badly worn surfaces" (Post 1915a, 26).

Post expressed being "idiotically excited" at crossing the Mississippi, as "two seeming educated and otherwise intelligent New York women told me that the middle of the United States was cut unbridgeably in two by the Mississippi!" (Post 1915b, 21). They made some headway across Iowa until rain turned decent roads into "the consistency of wet soap ... Without disputing any other territory's claim, we are willing to believe that all the mud championship medals of the world should go to Iowa" (Post 1915b, 21). Travel was miserable until the rains stopped and a day or two of sunshine dried them out and the roads were transformed "into good ones again" (Post 1915b, 25). With good weather, Post found the landscape of Iowa and the Midwest to be picturesque, comparing the spatial arrangement of farms to that of New England (Post 1915b, 26).

Their European motor car's clearance was too low for the heavily rutted poor roads in the American West: "Between Las Vegas and Santa Fe the going was the worst yet. Washed-out roads, arroyos, rocky stretches, and nubbly hills. We just about smashed everything. Cracked and broke the exhaust, lost bolts and screws, and scraped along the pan all of the way" (Post 1915c, 19). They limped their way to Winslow, Arizona and put their poor car on a freight train, taking the train themselves up to the Grand Canyon and then on to Los Angeles. Once the car was repaired, they resumed their tour. Eventually Emily's party reached their destination, arriving in San Francisco on June 8th, 1915. She concludes with a description of the Panama-Pacific Exposition, the beautiful architecture, the interesting displays, including airplane stunt flying exhibitions.

A year later, an expanded account of the trip was published in book form, complete with notes at the end "To those who think of following in our tracks" (1916/2004). Emily provides not only overall suggestions, but Ned provided comments on driving and repairs, there are recommendations on packing and dress, and at the very end of the volume are maps, a general US map with their route and then twenty-seven rough sketch maps capturing each stage of their journey with notes about where to stay and what to do (Figure 2.4). Emily took such detailed notes that, in her book, she provides her expenses for each leg of the trip, corresponding to the maps, including tips.

Through published accounts, such as Emily's, and media coverage of cross-country trips, such as Alice Ramsey's and the various suffrage tours, American women were being shown not only can women drive but where to

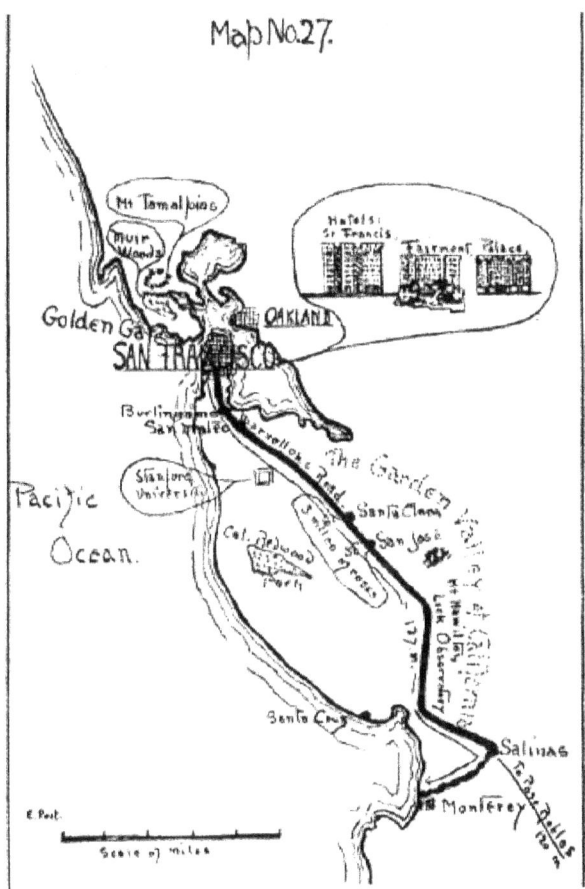

Figure 2.4 Example of map from Emily Post's *By Motor to the Golden Gate* (1916). This map depicts the last leg of their trip, from Monterey to San Francisco. Note the "E. Post" credit down above the scale in the lower left. While it is suspected that Emily was the creator of the maps, her son Edwin (Ned) Post cannot be ruled out.

drive and how. Unlike the bicycle where written accounts provided more local information about where to ride, descriptions of automobile touring emphasized distances, from living in commuter communities to traveling across
the country. The accounts emphasized not only the need to know how to
maintain and repair your vehicle but demonstrated that this was a task that
women could easily accomplish. By their example, women like Emily showed
other women how to conduct autotours, how to plan the trips and obtain
the information they needed, how to pack (and what not to pack!), how to
address the problems that developed along the way, but importantly was the
pleasure of motoring. There was a great deal about the beauty that could be
found in the American landscape, the history present, the scenic vistas, but
there was also the joy that came from motoring through the landscape.

Managing spatial information

In the early twentieth century, road information was spotty ... as were the roads.
In the Eastern states, *Blue Books* and other guides provided decent information
on road conditions and how to get from one major city to another (although
Alice Ramsey found them not entirely accurate). But in the South and West,
even less information was available. Women preparing for and in the process of
crossing the country had to assemble a variety of spatial information, as Emily
Post did. In 1906, Nancy Minerva Teape and her daughter Vera Marie Teape
traveled from Chicago to Denver and back.[33] In eastern Iowa, two elderly men
were concerned that they might lose their way and from their buggy pulled a
country map, "He came up to us and with his pencil showed us which road we
should follow. He explained that it was a very poor map and started in correct-
ing it. He placed several white schoolhouses on the other side of the road; also
removed bridges, placed principal farmhouses, etc." (Teape 1980, 6).

Magazines provide instructions on how to manage all these geographic directions. In a 1905 *Outing* article "Camping out with an automobile," the author
Hrolf Wisby begins with directions for creating a road map: "The first thing
to do is to select your route carefully. The chart will not tell you everything"
(Wisby 1905, 739). Wisby goes on to direct his readers to assemble the available
maps then to use tracing paper to create your own map composed of information from a "good, plain geographic map," a topographic map, and a survey
map. Once all the data has been traced onto the map, Wisby directs readers
to paste it onto canvas and rolled onto a cylinder, creating a strip map (akin
to an Automobile Association of American triptik). It is only *after* this that he
writes: "Next comes the choice of a proper vehicle... ." (Wisby 1905, 740).

Maps were so essential to the automobiling experience that in a 1906
poem entitled "The ABC of the Automobile," the letter M reads:

> M is the Model you choose with great care,
> The Map that you follow for roads that aren't there

Weeks 1906, 687

In some ways it is striking the number of references to poor maps, poor *Blue Book* directions, and roads that aren't even recognizable.

There is underlying all of these accounts an assumption that women can read and use maps. Never were there directions on how to *use* a map, just directions on where to get maps and how to manage them. *The Ladies' Home Journal* featured a "Puzzle Page" where readers who solved that month's puzzle could send in their answers in hope of winning a $25 prize. In July 1906, the puzzle was "A Unique Automobile Run": "A party of prominent men are contemplating a automobile trip from Washington to Chicago. Assuming that the route shown on the map below are the only ones between points, and that the arbitrary distances given are correct, what is the shortest mileage they would have to cover to visit every city shown, *entering each city only once?*" (Rosenfeld 1906, 35). The accompanying map covers the United States east of the Mississippi and features twenty-three cities with curvaceous roads running between. It is remarkable that women readers are given the challenge of reading the map and setting the course for the men.

The Good Roads movement and women's groups

> Blind curves, dangerous crossings and steep hills have wrecked many vehicles of transportation and cost many lives, especially in this day of somewhat reckless automobile driving. The Good Roads Committee of the Civic League of New Canaan, Connecticut, recognized this fact and set to work to improve conditions. It has cut down undergrowth and leveled hills, and set up markers so that travelers may be warned of dangers ahead.
>
> Wood 1913, 28

This quote, from the "What Women's Clubs are Doing" column in the *Ladies' Home Journal*, captures the actions of a women's club committee on improving the local roads. Women's clubs and organizations across the country took up the Good Roads cause and worked to improve the infrastructure in their communities. Their work on improving the roads and marking highways transformed the American landscape. Because this was done on a city-by-city, state-by-state basis, not all women's clubs chose to take on the Good Roads cause. I will highlight three well-documented examples: the work of Alma Rittenberry in Alabama, the marking of the Santa Fe Trail in Kansas, and the Boone's Lick Road and Santa Fe Trail in Missouri.

As mentioned previously, the Good Roads movement had begun by the League of American Wheelmen in the 1880s, with momentum gaining as the number of cyclists increased. With the invention and adoption of the automobile, Good Roads advocates shifted from being cyclists to being motorists. The American Automobile Association, founded in 1902, also advocated for better roads. In general, the Good Roads movement was calling for level, well-surfaced, low-dust roads, wide enough to accommodate both farm wagons and these new modes of transportation (Olliff 2010, 86). Fairly early on, Good Roads advocates sought to appeal to a wide constituency, appealing to rural Americans and their needs to get products to market (Potter 1891). As there was no federal agency that really addresses roads and their conditions at this point in history, "Americans fought over who would build and finance roads, what purpose roads served, and what place automobiles actually had in the American transportation system" (Olliff 2010, 89).

From publicized cross-country trips such as Alice Ramsey's had come an awareness of national road conditions and the lack of a "trunk line" linking major commercial centers (Olliff 2010, 89). Out of this developed the concept for the Lincoln Highway Association and other trail or road associations, such as the Yellowstone Trail and the Ozark Trails Association (Akerman 2002, 179). Often composed of manufacturers of automobiles and automobile parts, such associations sought to mark and improve major interstate networks, promoting automobiling and, of course, the prosperity of their own companies (Akerman 1993, 13). With the need for better roads, came also the need for *well-marked* roads. Named routes were "blazed" to mark the route, with markings usually consisting of painted color bands on telegraph poles; where routes intersected or overlapped, telegraph poles might be "adorned nearly top to bottom with various coloured marks representing the routes of different promoters" (Medlicott and Heffernan 2004, 243). Without a systematic road numbers system, early motorists depended on marked trails as well as the detailed directions published in *Blue Books* to navigate. But not all the organizations were tied to commercial interests: a significant number of women and women's organizations became involved with the Better Roads movement.

Miss Alma Rittenberry of Birmingham Alabama was instrumental in the development of the Jackson Highway, "an interstate road generally following the route General Andrew Jackson took to fight the Creek Nation and the British in 1813 and 1814" (Olliff 2010, 90). While Jackson's route began in Nashville, passing through Alabama to the Florida panhandle, then turning west to New Orleans, Rittenberry proposed a trunk line running from Chicago to New Orleans, working with the Alabama Daughters of 1812, a volunteer women's organization dedicated to perpetuating the memory of the War of 1812. Observing that there was no memorial to Andrew Jackson in Alabama, they decided to commemorate him with a highway (Rittenberry 1916). Rittenberry chaired the committee and began her work in spring 1911, endorsed by the National Good Roads Congress. At the time, there was no federal authority in charge of interstate highways, so Rittenberry concentrated on local governments and counties: "To excite interest and local support, she traveled the entire proposed route, speaking before commercial bodies, chambers of commerce, women's clubs, civic groups, and local and state good roads conventions" (Olliff 2010, 95). While some of these presentations were likely before small audiences, not all were: in 1912, she spoke before an audience of 2500 ("Interest in ..." 1912). Rittenberry also published numerous articles in magazines, including multiple articles in *Southern Good Roads*. In promoting the Jackson Highway, Rittenberry emphasized the benefits of better roads for all but especially for rural families (Interest in … 1912, 27).[34] As automobile touring grew, she added the economic benefits of tourism to her argument. Rittenberry worked tirelessly promoting the Jackson Highway, largely at her own expense.

In 1915, she organized "a multistate trail group she dubbed the Jackson Highway Association," hoping that a broader-based group would lead to greater advancement of the concept (Olliff 2010, 97). But with the Association came loss of control. Other states, Mississippi in particular, lobbied for an alternative route through their state. In a hostile meeting in 1917, the Association abandoned "Rittenberry's careful work and her dream of raising a monument to Jackson in Alabama as well as efficiently tying the state's commercial centers together" (Olliff 2010, 100). Rittenberry resigned from the Association, taking the Alabama delegation with her and forming the "North-South National Bee Line Highway Association," heavily weighing the board with Alabamans. The Bee Line would not only be a trunk line through Alabama, they extended it to north to Nashville and south to Florida. Eventually, they agreed to connect with the Dixie Bee Line Highway, connecting Nashville through Terre Haute to Chicago and southeast to Tampa.

World War I slowed progress on the Bee Line. Momentum was lost and the interest in associations promoting such highways declined. U.S. Congress finally took up the issue of interstate highways. While Rittenberry's efforts did not result in her dream of a Jackson Highway through Alabama, and the Bee Line had limited success as part of a federal highway system, Rittenberry's efforts were at the heart of both and she did tremendous work through her public speaking and her publications to bring greater attention to the need of good roads in the South.

The Daughters of the American Revolution (DAR), a women's service organization composed of descendants of Revolutionary War patriots, began a nation-wide project in 1912 to build a national highway, through their National Old Trails Road Committee (NOTRC) (Medlicott and Heffernan 2004, 234). DAR was founded in 1890 and spread rapidly throughout the United States. DAR activities usually concentrated on historic preservation, education, and promoting patriotism (Medlicott and Heffernan 2004, 236). Members in Western states were initially challenged with historic preservation, but quickly turned to preserving more recent history: that of the American frontier. The Kansas DAR were the first to take up the challenge, focusing on the Santa Fe Trail in 1902 (Cordry 1915; Medlicott and Heffernan 2004, 238; Stanley 1907). The Santa Fe Trail was a major trail through central United States, connecting Franklin, Missouri with Santa Fe, New Mexico. From the 1790s till the road reached Santa Fe in 1880, it was the commercial route with both freight and immigrants following its path across the southern Plains. The Kansas DAR took on the service project of marking the Trail that was rapidly being forgotten: "All traces of the trail had been obliterated or nearly so and we were compelled to enter upon a study of the subject and acquire information by all possible methods, and sometimes such information was difficult to procure" (Stanley 1907, 65). In their process of marking the trail, they had to establish the exact route of the trail, engaging with archival and qualitative research (interviews) in order to establish the spatial extent of the Trail and place 96 markers so that it would not be forgotten (Figure 2.5).

The Kansas DAR's work progressed slowly at first, hampered by the "difficulty in getting a true map" of the Trail (Cordry 1915, 21). Eventually, with the aid of the Historical Society, they secured "the map of the Trail made by the Government in 1825–27, the recently made map of the U.S. Government, topographic maps, which cover the route, and upon which it was supposed the Trail could be drawn with sufficient minuteness to enable persons on the ground to find the Trail with comparative ease" (Cordry 1915, 23). Roy Marsh of Topeka was hired to compile their map and they began to raise funds to place markers along the trail. As word spread about the project, they began to receive letters from Kansans who had either traveled the Trail or lived along it, supporting the project and telling their stories. The greatest uncertainty came to the exact route west of the "great bend of the Arkansas river. The fact that the counties in western Kansas were very large when the last topographic map was made by the Government, and that they had since been divided, thus changing county and section lines, made the work all the harder" (Cordry 1915, 39). As part of an effort to promote the effort and raise funds for the markers, a small version of the map (4 × 8 inches) was also created by Marsh (Cordry 1915, 52).

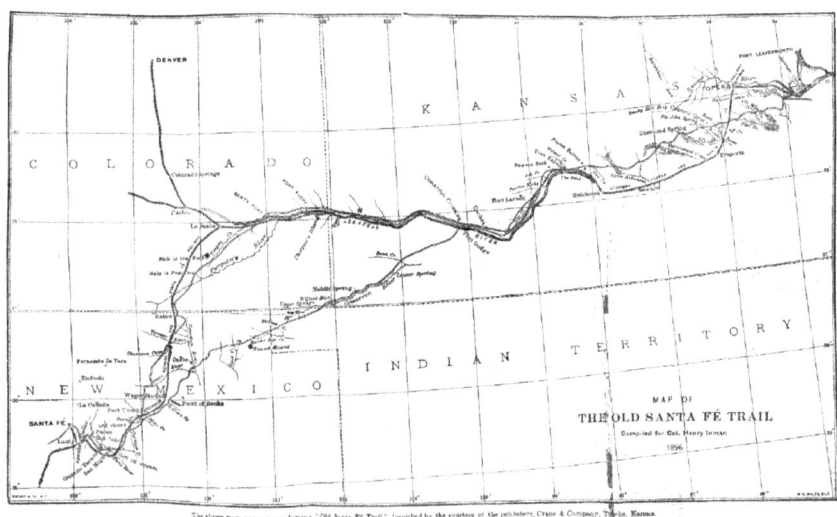

Figure 2.5 Map of the "Old Santa Fe Trail" from A. Cordry, *The Story of the Marking of the Santa Fe Trail by the Daughters of the American Revolution in Kansas and the State of Kansas* (1915). Despite appearing in the DAR account of finding and marking the trail, this does not appear to be the map that they created but rather an existing published map.

As work on the map progressed, the Kansas DAR enlisted county cit-izens and clubs to help with fundraising and to designate locations along the Trail for the markers. Kansas school children were involved with the project, with an essay contest held (Stanley 1907, 65). The philosophy of the Kansas DAR on markers was that they would place as many as they could afford: the more money raised, the more markers would be placed. Cere-monies were held in each community to dedicate the markers. The granite markers identify the trail and approximate dates of use, then names the DAR and the state of Kansas as sponsors. A "prominent newspaper man and politician" questioned DAR being named first, to which he was told "Why, man, the State of Kansas has had this Trail for nearly fifty years, and lost it,—the Daughters of the American Revolution in Kansas found it" (Cordry 1915, 16).

The Kansas DAR's work to correctly map the Santa Fe Trail and then provide visible markers along its route represents women's construction of landscape. They conducted research to determine the correct route and hired someone to compile the results. They debated the construction of the markers (concrete versus granite) and the appropriate text. They scrambled to raise funds for the markers. Through their efforts, they were materially shaping their local landscapes.

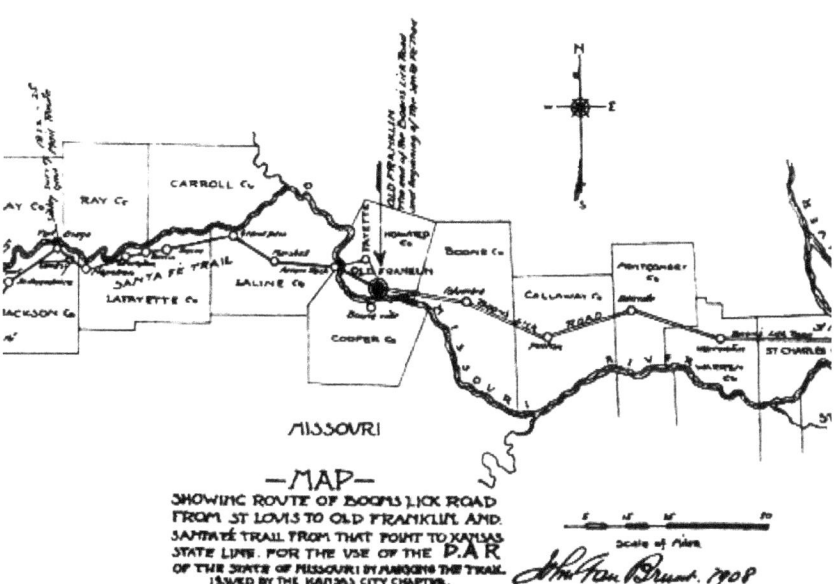

Figure 2.6 "Map showing route of Boon's Lick Road from St Louis to Old Franklin, and Santa Fé Trail from that point to Kansas state line ...," drawn by John Van Brunt. From E. Gentry, "The Boone's Lick Road and Santa Fé Trail: The Missouri Cross-State Highway" *The American Monthly Mag-azine* 39, 4 (Oct. 1911): 186.

The work in Kansas led to the marking of the Santa Fe Trail in Colorado and Missouri (Cordry 1915, 142). The Missouri DAR picked up the challenge of preserving the Trail in Missouri and took it a step further. They merged Santa Fe Trail preservation with a "Good Roads" committee that had existed since 1907 and proposed constructing two roads on historic trails to create a trans-Missouri highway, "that the old trails of the state should be reblazed into modern roads as a monument to the pioneers of the state" (Medlicott and Heffernan 2004, 240; Woman Road Builders 1912) (Figure 2.6). The Missouri plan was initiated and headed by Miss Elizabeth Gentry Butler. To promote their plan, "the Missouri Argonauts of to-day set sail across the State … in fifty motor cars seeking the best route for a cross-State highway (Gentry 1911, 187). The "expedition" was headed by the Missouri governor, members of the State Board of Agriculture, and Santa Fe Trail Committee members of the Missouri DAR including Gentry, "inspecting" each segment of the route and enlisting support from local commercial clubs and businesses. The Missouri DAR members had ambitious plans, for they envisioned their highway as a component of a national system based on old trails that they would call "The DAR National Old Trails Road" (Figure 2.7).

As part of their campaign materials, they created maps of the route, at the regional scale and at the national: "the route as mapped by Mrs. John Van Brunt is the most practical route, the most scenic, and the most historic of any route suggested as a transcontinental highway" (Gentry 1913, 161). While Mrs. Van Brunt may have "mapped" the route, the maps were drawn by her husband John. The national map became a visual touchstone of their campaign, appearing not only in D.A.R. publications but also in more widely circulated publications such as *Twentieth Century Magazine* (Woman Road Builders 1912).[35]

Missouri DAR convinced a Missouri congressman to introduce a bill into Congress in 1911 which would have provided federal monies to the states for road improvements (Medlicott and Heffernan 2004, 246). Unfortunately, the bill did not pass. Through their instigation, a men's organization the National Old Trails Road Association was created, becoming "the centerpiece of what had long been a thriving good roads movement among politically minded men in Missouri" (Medlicott and Heffernan 2004, 247). The NOTR Committee designed a marker that they hoped to use to mark every one-mile interval of the Highway. They sold "Madonna of the Trail" pins and DAR pennants to fund the production of the mile markers (Gentry 1914). In the meantime, women marked the trail themselves with red, white and blue paint on telephone poles (Gentry 1913, 161; Medlicott and Heffernan 2004, 251). Through the painting of the telephone and telegraph poles, the route was marked across the states. While Congressional monies never came through, the work of Gentry and DAR accomplished a great deal.

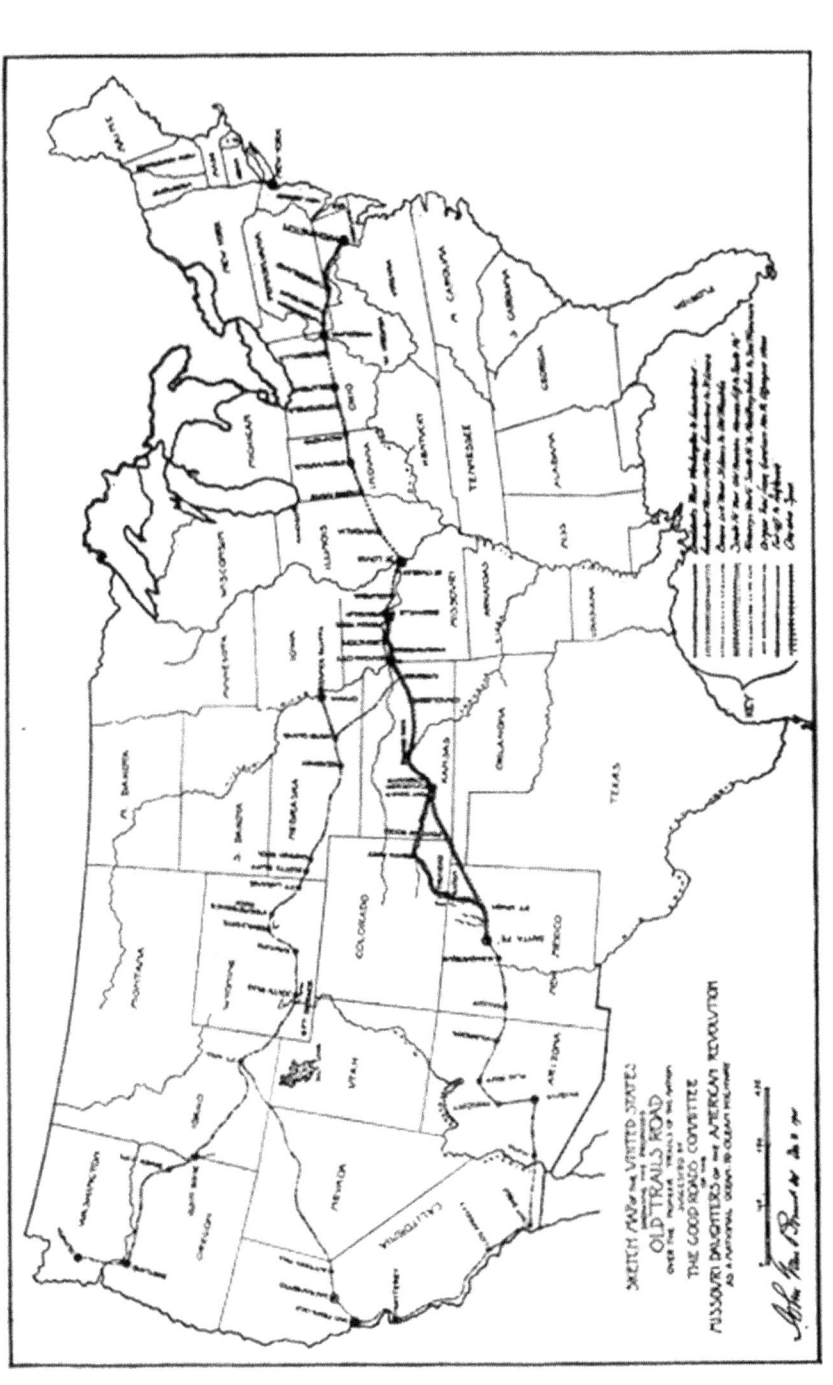

Figure 2.7 "Sketch map of the United States showing the proposed Old Trails Road over the pioneer trails of the nation suggested by the Good Roads Committee of the Missouri Daughters of the American Revolution as a national ocean-to-ocean highway," drawn by John Van Brunt, from E. Gentry's "The Old Trails Road," *The American Monthly Magazine* 42, 4 (April 1913): 162.

Associations began to lose their hold as Congress began to take the lead in road improvement efforts, beginning with the Federal Aid Road Act of 1916 and extending it with the Federal Aid Road Act of 1921 (Olliff 2010, 105–6). In 1924, the American Association of State and Highway Officials (AASHO) "petitioned to the U.S. Department of Agriculture through its Bureau of Public Roads to establish a more systematic means of marking highways than allowing trail associations to mark their own" (Olliff 2010, 107). Through AASHO, highways were approved for an interstate network and a systematic numbering system created.

The work of Rittenberry and the Alabama Daughters of 1812 and the state and national Daughters of the American Revolution on roads was but the tip of an iceberg. In gathering data on these efforts, I inadvertently came across other efforts by women to improve roads. On the same page as an update of the Jackson Memorial Highway, was an account of a woman in Kansas who had maintained the road in front of her house for the past nine years. Miss Betty Galwith stated:

> I drag the roads because I believe in good highways... . My father always took good care of the roads in front of his land and I am just continuing the work since his death. So many people around here think it is queer that I should get out and work on the roads, but I don't. You see I have been doing it so long now.
>
> Woman Drags Road 1912

In gathering data on Alma Rittenberry, I ordered by Interlibrary Loan an article that I thought was about her entitled "The Grandmother of Good Roads." It was not about Rittenberry, it was about Anna Norris Kendall of Deer Park, Alabama, who at the age of seventy-nine was a road commissioner in her county and could be found "standing bareheaded by the side of the road, directing the driver and his team of mules building the new highway" (Norris 1924). While working to improve her county roads, she was also deeply involved in the "Mississippi Valley Highway" to run from Duluth to the Gulf and the construction of a much needed tri-state bridge at Cairo IL. It is clear that *all* the women who worked on improving the roads believed in the importance of their work and were proud of their accomplishments in contributing to the American landscape.

American women and automobility

With the adoption of the automobile, American women saw a significant transformation in their landscapes. Between 1910 and 1920, automobiling shifted from a pastime of the wealthy to a pastime of the middle class, aided by the availability of less expensive cars like the Model T (Preston 1991, 39). With greater availability of both new and used automobiles, more Americans were able to take advantage of its benefits. The automobile expanded the zone of mobility for many women who had been previously restricted, be it by lack of transportation or by poor roads. A woman commented on the DAR's Old Trails Road that "My men folks have left me stuck in the mud all my life. I am mighty thankful the D.A.R. are trying to pull me out" (Gentry 1913, 163). Christine Frederick, writing in 1912 about the life of a "commuter's wife," describes how the automobile "annihilates" barriers, with those barriers being both spatial and societal (Frederick 1912, 14). With the automobile and with better roads, they could explore new places with ease, seeing their own country for the first time, or just go to see family or friends a short distance away.

As women traveled in their automobiles, they were transcending limitations and stereotypes. Social and geographic restrictions on where women could be grew less. Stereotypes about the "women's sphere" and women's abilities appeared less often as women demonstrated that they not only could drive their own cars but also repair them and take on cross-country trips over dubious roads (Wosk 2001, 115). Fixed divisions between the sexes began to shift (Clarsen 2008, 8). There were still complex and variegated social relations when it came to driving, but things were markedly improving.

As the spaces women occupied expanded, new information was needed about how to efficiently move from one place to another. By publishing accounts detailing the problems with insufficiently detailed maps or incorrect *Blue Books*, women were bringing the public's attention to the need for better maps and for better roads. Sometimes, as in the work of the Kansas and Missouri DAR chapters, they were reclaiming trails that were nearly forgotten, remapping them, and even sometimes turning them into part of the highway network.

As they moved through the American landscape, American women were encountering their own country in new ways. Emily Post, who had traveled extensively in Europe, was ecstatic in discovering the beauty of the American landscape. However, this experience of the landscape was an interesting mixture of independence and constriction. Many of the accounts by women emphasize the freedom offered by automobiling, of being able to go where you want, when you want. Yet, it is clear that not all the roads were appropriate for automobiles and that road conditions restricted where you could safely drive. And there is the narrow view of landscapes provided by automobiling, of limited corridors through the landscape that the autos speed along, with the driver's gaze directed through the front windshield. American women were seeing their country in new ways.

The expansion of mobility through automobiling also resulted in the creation of new spaces. The corridors of consumption that began with the bicycle, expanded dramatically with the automobile. The more Americans drove, the more they consumed (Gudis 2010, 371). American women helped to create these corridors through their work on the national road system, whether it was simply dragging the road in front of their home or working on the concept of a transnational highway. This development of an interstate road system was essential in linking our nation. Emily Post describes how when she was planning her trip across the country, she obtained a national map: "a large one of the United States with four routes crossing it, equally black and straight and alluring" (Post 1915a, 11). It is hard today to image that there were only four routes across the country, and questionable routes at that. By working to improve the nation's road and advocating for good roads, American women were both improving their own mobility as well as contributing to the development of our nation. America was an empire but not a very well-woven together one. By promoting and even physically developing roads (dragging in front of their farm, serving as a road commissioner and overseeing road development), American women were contributing to the infrastructure necessary to integrate the country.

In addition to physically connecting our country, the East with the rest, the automobile and autotouring also facilitated a "conceptual ownership" of the landscape by making it possible for more Americans to see and experience their country, colonizing and commodifying what lay beyond the windshield (Gudis 2010, 373). In the early 1900s, we were still adding states to our union. There were individuals still homesteading, still virgin lands being put under the plow. Automobiling allowed us to complete our manifest destiny. While technically we had all the territory, we were still making it our own. Through construction of roads and infrastructure, as well as facilitating more Americans to travel and see and identify with the American landscape, we were extending our geographical imaginations to the whole of our own country.

Mobility and geographic knowledge

Margaret Le Long, Frances Willard, Alice Ramsey, and Emily Post all represent the forefront of American women who were able to take advantage of greater mobility in the Progressive Era. Not all American women were able to do so, but many did after them, over the following century. Alice Ramsey would continue to make transcontinental trips almost yearly: "She stopped counting after her 30th transcontinental trip" (McConnell 2000, 59).[36] But it was not just being able to move yourself freely from one place to another. In 1895, Jean Porter Rudd captured the spectrum of mobility:

> [A] wheel is not merely a conveyance, a vehicle. It is a whole code of philosophy, it is the world, the universe, the much in little.
>
> Rudd 1895, 127

In the 1890s, many women cyclist/writers equated mastering the wheel with mastering life, such as Frances Willard. But mobility provided to be a transformation experience for American women and even for the American landscape.

With greater mobility came new frontiers. As Americans moved around their landscapes in new ways, first with the bicycle, then with the automobile, came the need for wayfinding knowledge, knowledge not only of where you should ride (types of surfaces, safe and/or appropriate locations) but also the spatial distribution of these spaces. The automobile transcended greater distances and, as a result, required even more geographic knowledge to accompany these distances. Women shared knowledge of where to ride with each other, writing articles and books encouraging other women that they too could break the bonds of immobility and make their way in the world. Practical geographic knowledge was circulated more widely than it ever had before.

As new mobilities were adopted by Americans, new geographies were created. Not only were there new ways to move through space and across landscapes, but there were new spaces created for these mobilities. Bicycles began the process of road improvement in the United States. With the automobile, there was even greater need for not only improved roads but interstate road networks connecting our country. While some roads were built on existing roads, or even on old trails, new highways were also created. As Americans used particular roads, such as for tourism in New England, or long distance travel along the Lincoln Highway, corridors of consumption developed as businesses located along these routes, hoping to attract customers. New businesses developed to serve these consumers, such as bicycle repair shops, gas stations, auto repair shops, restaurants and hotels. As new "trails were blazed," maps and other forms of geographic wayfinding information (such as *Blue Books*) had to be updated and corrected to accurately convey changing conditions, resulting in another industry to benefit from the mobility boom.

As American women became more mobile, their place in the American landscape changed literally and figuratively. Certain classes of women were always mobile, walking to market or to work, riding horseback to teach school or visit a neighbor, taking a wagon to pick up supplies. Sometimes, their mobility was not of their own choosing. As historian Virginia Scharff has observed, we often forget that uneven power relations, such as patriarchy, results in women being obliged to move whether they want to or not when their husbands/partners/fathers make the decision to move, such as following the Oregon Trail or emigrating to the United States. Yet despite their secondary role, women's knowledge and work was frequently crucial to the success of the venture, making it possible for the families to move and keeping the families functional: "Every time they washed a diaper in a creek or threw another buffalo chip on the fire, they transformed their immediate environs, and made travel the next day possible. Women's knowledge, labor, and motion were thus critical to inscribing the trail across the continent" (Scharff 1999, 294).

With the bicycle, women entered the landscape in new ways. Some women did use the bicycle for practical purposes, such as during the Philadelphia street car strike of 1895 when women who worked in the business district used them to get to work or to go shopping. A significant change was women's use of the bicycle for leisure, for pleasure. Again, this was not new: women would walk or ride a horse to enjoy a nice day. With the bicycle, it was giving women who may not have the means to own a horse, some of the pleasures of horseback riding – the sense of speed, the wind in your hair, being able to go farther, faster. But unlike the horse, the bicycle was powered by its rider – adding to the pleasure was the knowledge that this was your own doing, you were driving it, you were propelling it along, it was all you.

This new placement of women in American landscape had its benefits and its drawbacks. For many women, the bicycle then the automobile opened new worlds to them. It certainly began a revolution in women's dress which resulted in greater bodily mobility whether you were on a bicycle or not. Women experienced environment and landscape in new ways and offered women freedoms they had previously not experienced. On the other hand, the presence of women bicycling in the landscape was viewed as startling, largely tied to concerns of dress and behavior. Eventually, women demonstrated that they could ride bicycles, be dressed appropriately, and behave in a lady-like manner. Women bicycling became a normal sight, even celebrated on a lovely day. Similarly, women driving automobiles was seen initially as such an unlikely event that it drew crowds. This too eventually became a regular occurrence. The presence of women in public spaces, whether it was working in some capacity or in transit from one place to another, became a part of the new American landscape.

As much as women were now seen as part of the landscape, there were still moments when they perceived to be "out of place." In general, women who moved too freely, whether on a bicycle without a corset or too fast in an automobile, bucked gender conventions, "becoming actual forces of disruption and potential agents of historical transformation" (Scharff 1999, 294). To be "out of place," can be literal or figurative. Where they were riding, be it city streets or country roads could be viewed as being in inappropriate territory for a woman. Women riding alone, unchaperoned, was crossing a line socially. Whether bicycling or automobiling, women became subject to the public's gaze on all aspects of their being – dress, driving behavior, language – and subject to "regimentation, policing and surveillance" (Gudis 2010, 373).

Despite women coming to cycling and motoring after men, in women's literature they are presented as being better than men at these tasks. In an anonymous essay entitled "Where Women Motorists Excel Mere Men," the author extolls the virtues of women drivers, especially during World War I, where they did better on driving exams, were adept at diagnosing engine problems, and that "The fact that a woman makes a good mechanic or driver does not in the least mean she is not equally apt at housework or cooking" (Where Women Motorist … 1920, 76).

Several articles discuss women driving their husbands to work or to the train station (Frederick 1912; Martin 1911). An article on women motoring in Detroit states that by acting as "chauffeuses" to their husbands, Detroit wives keep track of their husbands, "driving them to their places of business in the morning and calling for them when the day is done. More than one Detroit woman makes a practice of calling for her husband, with her electric-car, at his club, when it is time for him to go home" (Martin 1911, 12). And it has been suggested that they are also better at map reading and reading the landscape. While women were culturally not associated with mobility, there is the opinion (at least in the women's press), that they are better suited to the tasks associated with mobility – driving and wayfinding.

As women made their way through the American landscape, they saw areas which could be improved, in terms of roads and in geographic information. American women not only belonged to the American landscape, they made a visible, material difference, contributing to the building of America, improving their country. They published directions on how to travel by bicycle and by car. They extolled the virtues of touring and seeing especially the American landscape. Both individually and through a variety of organizations they began to campaign for improvements to the nation's roads. Through their presentations and publications they made the case for the development of the nation's interstate highways and how it would benefit *all* Americans, not just those bicycling or automobiling.

Through their bicycling and automobiling, American women contributed to the development of American geography. This geography was both the physical landscape (roads) and the geographical imaginations of Americans. Through their published accounts of bicycling and automobiling, they were filling in aspects of the American landscape that their readers may not previously considered. Whether they bicycled or automobile or not, the information provided through their accounts added to the geographic imaginations of their readers. The women writers of such accounts may not have been professional geographers, yet they were contributing to the geographic knowledge circulating among the American public.

Notes

1 Le Long also found the map of Wyoming misleading: "To look at the map of Wyoming, one would think it a thickly-settled country. Had I known how many of those names that loom so large on the map stand only for side-tracks and section-houses, I fear I should never have had the courage to make the trip" (Le Long 1989b, 595).
2 There is also the airplane, of course. Women were also early enthusiasts of flight but not to the extent of the bicycle and the automobile. For the purposes of my research, I concentrate on the means of transportation most widely available and adopted – the bicycle and automobile.
3 Republished in 1991 as *How I Learned to Ride the Bicycle: Reflections of an Influential Nineteenth Century Woman*.
4 Petty estimates that "there were probably fewer than 100 regular women tricyclists in the U.S." (Petty 1997, 117).
5 An 1888 *Godey's* article on "Tricycling" includes a discussion of what to wear when you are tricycling (Dodge 1888, 210). And in 1890, Godey's included a line drawing and a description of a tricycling costume in its Fashion section.

6 Columbia advertisement appeared in: *Godey's* April 1894, 18.

7 There was also public debate on the question of whether it was medically appropriate for women to ride bicycles. The most significant concern appears to have been the use of a straddling seat and the potential of women getting sexually excited by cycling. This is beyond the scope of my research. But for overviews of this research, see Garvey (1995) and Macy (2011b).

8 A wedding gift from a beloved uncle was a tandem bicycle, his and hers cycling outfits, and two tickets to New Orleans (Follett 1896, 5).

9 Smith unfortunately does not provide a reference for this Chicago episode.

10 This includes Black Americans. It is extremely difficult to get meaningful numbers on the number of riders in general, let alone minorities. But anecdotes and photographs provide evidence of Black American ridership. One of the most famous bicycle racers of the 1890s was Marshall Taylor, a Black American from Indiana. And there was a Black bicycle corps in the Army based out of Missoula, Montana (The 25th Infantry Bicycle Corps). On the downside, there are many cartoons ridiculing black bicyclists from the 1890s. For more information, see: Ritchie (2003); Foster (1999); and Koelle (2010). Lorenz Finison's *Boston's Cycling Craze, 1880–1900* provides a fascinating study of the great variety of cycling clubs in the Boston area during the peak of the cycling craze, in particular documenting Black American cyclists and cycling clubs as well as evidence of other minority cyclists in the Boston area (Finison 2014).

11 See for example H.G. Wells' novel the *Wheels of Chance: A Holiday Adventure* (1896); Kate Chopin's short story "The Unexpected" (1895); and Willa Cather's short story "Tommy the Unsentimental" (1896).

12 "Daisy" was written by Harry Dacre in 1892.

13 Usually, but not always, advocating. There were published arguments against women's cycling, such as "Crusade Against the Wheel for Women" (1896).

14 Emphasis is from Mackintosh and Norcliffe (2007).

15 "The World Awheel" describes an exclusive New York bicycle club with indoor space, the Michaux Club. Townsend also describes the Michaux Club but goes on to describe "its imitators in Brooklyn, Richmond, Va., Philadelphia, Chicago, Louisville and other cities." Spreng describes how in 1895 in Grand Forks ND "Entrepreneurs converted indoor floor space into 'bicycle academies' where novices could take their first falls" (1995).

16 In Finison, see especially Chapter 1.

17 In the back of the *Wisconsin Tour and Hand Book*, is essentially a list of why bicyclists should become members, including: "Join the League and aid in the support of the great movement for the building of permanent highways" and "Do you ride a wheel? If so, you are a loser unless you are a member of the Wisconsin Division, L.A.W." (1897, 116).

18 Alice Moqué describes running into an old acquaintance at a church in Stratford-on-Avon: " ... for in the rector we recognized an old friend, he having lectured at the Geographical Society in America for the benefit of the American memorial-window, to be placed in this very church, the burial-place of the 'Bard of Avon'" (Moqué 1897, 460).

19 Ramsey eventually publishing her account in *Veil, Duster and Tire Iron* (1961), republished in 2005 as *Alice's Drive*. Her papers are held at Vassar College and include a scrapbook documenting the trip.

20 Describing one day's trip across Wyoming, Ramsey writes: "But as we followed the trail in open—very open—country, we came to a place where the road crossed the railroad at the top of a small rise. Beyond this point there was no road. It just plain quit" (Ramsey 2005, 100).

21 For an excellent overview on the Pope Manufacturing Company and its combination of industrial methods, technological advancements, and advertising
   innovations, see Epperson (2000).
22 The first automobile road map in the United States may have been a newspaper
   map showing the route of the race (Akerman 1993, 11–12).
23 Winston Motor Carriage ad appeared in *Scientific American* 79, 6 (August 6,
   1898): 96.
24 All the accounts I have examined, of women writers advocating automobiling, the women are driving gasoline automobiles (Irving 1909; McManus 1912;
   Hitchcock 1913; Murdock 1915, "Little Things ..." 1917, Ramsey 2005).
25 Minna Irving's account of learning to drive and care for her car, "My lady of the
   car," states as one point: "Having played piano was a distinct advantage to me in
   learning to drive. It is much easier for a pianist to become an expert automobilist
   than for one who has not already learned to be ambidextrous, and to use both
   hands, as well as the feet, independently" (Irving 1909, 234).
26 Oldsmobile advertisement appeared in *The Ladies' Home Journal* 22, 5 (1905): 52.
27 Waverly advertisement appeared in *The Ladies' Home Journal* 19, 3 (1902): 34.
28 For example, in *The Ladies' Home Journal* 24, 4 (1907): 50.
29 Emphasis is Ramsey's.
30 The American suffrage movement adopted the color yellow as part of its campaign, borrowing it from the Kansas suffrage campaign. This was used to dif-
   ferentiate the American movement from the British, who used the colors purple,
   white, and green (Sewall 2003, 91).
31 Later in life, Post would be well known for her books and columns on etiquette.
32 Interestingly, Post initially tells him that she'd like to take the Santa Fe Trail. There
   is no way to know if she was aware of the work being done by the Kansas Daughters
   of the American or the National Old Trails Road Committee of the DAR.
33 Vera and Nancy would attempt an even more ambitious trip in 1908, a cross-
   country tour from Portland ME to Portland OR. They would make it as far as
   Kansas City before Nancy's health put a stop to the trip.
34 Rittenberry in a letter to *American Motorist* magazine writes: "The farmer does not
   want a speedway. One old farmer told me, in one of my rounds, that if he had his way
   he would put a log across the road and ditch every automobile" (Rittenberry 1913).
35 The version of the map that appears in *Twentieth Century Magazine* is the same as
   in D.A.R. publications, however it is reproduced much smaller and is virtually illeg-
   ible. It is assumed that readers would just get a sense of the national network from
   the map as it is difficult to discern the city names and the different trail patterns.
36 I've not been able to find more information about Margaret LeLong and her life.
   But Alice Ramsey's story is better documented, thanks to the American Automobile Association recognizing Ramsey as "Woman Motorist of the Century"
   in 1960, leading to the publication of *Veil, Duster and Tire Iron* in 1961. Ramsey would resume her life as a New Jersey wife and mother after her adventure
   but continued to drive, even taking her children on cross-country trips. Ramsey
   lived to be 97 years old (McConnell 2000).

Primary Sources

Adams, S. 1913. Blue Mondays: what they are likely to mean to the average American woman. *Ladies' Home Journal* 30, 2: 20.
Bly, N. 1896. Champion of her sex. Miss Susan B. Anthony tells the story of her
   remarkable life to "Nellie Bly." *The World* 2 February, 10.

Brown, H. 1895. *Betsey Jane on Wheels; A Tale of the Bicycle Craze*. Chicago, IL: W.B. Conkey Company.

Bruce, R. 1900. The place of the automobile. *Outing* 37, 1: 65–69.

Cordry, A. 1915. *The Story of the Marking of the Santa Fe Trail by the Daughters of the American Revolution in Kansas and the State of Kansas*. Topeka, KS: Crane & Company, Printers.

Crusade against the wheel for women. 1896. *The Literary Digest* 13, 12: 361.

Davidson, L. 1896. *The Handbook for Lady Cyclists*. London: H. Nisbet and Co.

Denison, G. 1891. How we ride our wheels. *Outing* 19, 1: 52–54.

———. 1893a. Through erin awheel. *Outing* 22, 1: 28–34.

———. 1893b. Through erin awheel II. *Outing* 22, 2: 136–140.

———. 1893c. Through erin awheel III. *Outing* 22, 3: 236–240.

———. 1893d. Through erin awheel IV. *Outing* 22, 4: 311–315.

———. 1893e. Through erin awheel Concluded. *Outing* 22, 6: 447.

Dodge, C. 1888. Out-door athletics for American women: tricycling. *Godey's Lady's Book* CXVIII, 693: 208–210.

Erskine, F. 1897/2014. *Lady Cycling: What to Wear & How to Ride*. London: British Library Publishing Division.

The feminine scorcher. 1896. *Godey's* 32, 790: 446.

Follett, H. 1896. A honeymoon on wheels. *Outing* 29, 1: 3–7.

Frederick, C. 1912. The commuter's wife and the motor car. *Suburban Life* 15, 1 (July): 13–14 and 46.

Gentry, E. 1911. Boone's Lick Road and Santa Fe Trail: The Missouri cross-state highway. *The American Monthly Magazine* 39, 4: 185–187.

———. 1913. The Old Trails Road: proposed by the D.A.R. as a national ocean to ocean highway. *The American Monthly Magazine* 42, 4 (April): 161–165.

———. 1914. National Old Trails Road Department. *Daughters of the American Revolution Magazine* 45, 2 & 3 (August/September): 132–135.

Godfrey, A. 1898. Awheel over the Jersey Highlands to the sea. *Outing* 32, 1: 8–15.

Hitchcock, A. 1913. Woman at the motor wheel. *American Homes and Gardens* 10 (April): vi and viii.

H.T.P. 1901. The return of the horse. *The Bookman* 13: 425–426.

Humphreys, P. 1906. An automobile vacation on $1.60 a day. *The Ladies' Home Journal* 23, 8: 27.

Interest in Jackson Memorial Highway Growing. *Southern Good Roads* September 1912: 27.

Irving, M. 1909. My lady of the car: motoring as a recreation for women – a personal experience and a bit of romance. *Suburban Life* 9, 5: 233–234 and 280–281.

Julka, L. [1941]. *Winnebago Wheelmen, L.A.W. 400: Bicycling, 1891–1941*. [Fond du Lac? WI: L. Julka]

Levitt, D. 1909/2014. *The Woman and the Car: A Chatty Little Handbook for All Woman Who Motor or Who Want to Motor*. Oxford: Old House.

Le Long, M. 1898a. From Chicago to San Francisco awheel. *Outing* 31, 5: 492–497.

———. 1898b. Alone and awheel, from Chicago to San Francisco. *Outing* 31, 6: 592–596.

Little things about a car that every woman who means to drive ought to know. 1917. *The Ladies Home Journal* 34, 3 (March): 32.

"Martha." 1892. We girls awheel through Germany. *Outing* 20, 4: 298–302.

Martin, I.T. Where woman runs her car. *Harper's Weekly* May 1911, 12.

Mayor aids suffragists. *The New York Times* November 20, 1915.

McCulloch, C. 1910. How to conduct automobile trips. *The Woman's Journal* 41, 31 (July 30): 121.

Merington, M. 1895. Woman and the bicycle. *Scribner's Magazine* 17, 6: 702–704.

Moqué, A. 1896a. A Bohemian couple wheeling through western England. *Outing* 28, 3: 186–191.

———. 1896b. A Bohemian couple awheeling through west England. *Outing* 29, 3: 232–235.

———. 1897. A Bohemian couple a-wheeling through middle England. *Outing* 29, 5: 460–466.

Murdock, A. 1915. The girl who drives a car. *The Ladies' Home Journal* 32, 7 (July): 11.

Norris, M. 1924. The grandmother of good roads. *The American Magazine* 98 (August): 63.

Post, E. 1915a. By motor to the fair: a trip from New York to San Francisco. *Collier's Magazine* 55, 25 (September 4): 11–12 and 24–28.

———. 1915b. By motor to the fair: a trip from New York to San Francisco. *Collier's Magazine* 55, 26 (September 11): 20–21 and 25–26, 28 and 30.

———. 1915c. By motor to the fair: a trip from New York to San Francisco. *Collier's Magazine* 56, 1 (September 18): 18-19, 22-24, and 26-28.

———. 1916/2004. *By Motor to the Golden Gate.* Ed. J. Lancaster. Jefferson, NC/ London: McFarland & Company, Inc.

Potter, I. 1891. *The Gospel of Good Roads: A Letter to the American Farmer.* New York: The League of American Wheelmen.

———. 1896. The work of wheelmen for better roads. *Godey's Magazine* 132 (April): 349–354.

———. 1896. The bicycle's relation to good roads. *Harper's Weekly* 40 (April 11): 362.

"The Prowler." 1896. Cycling monthly record. *Outing* 28, 3 (June): 42–48.

———. 1898. Cycle touring. *Outing* 32, 3: 320–324.

Rittenberry, A. 1913. The Jackson Highway. *American Motorist* 5, 2 (February): 142.

———. 1916. The Jackson Highway. *Southern Good Roads* April 1916: 11–13.

Road Book of Massachusetts: containing also some of the principal through routes of other states and Canada. 1898. Boston: League of American Wheelmen.

Rosenfeld, H. 1906. A new puzzle page: a unique automobile run. *The Ladies' Home Journal* 23, 8: 35.

Rudd, J. 1895. My wheel and I. *Outing* 26, 2: 124–128.

Stanley, W. 1907. Marking the Santa Fe Trail. *Santa Fe Employes' Magazine* 1, 3 (February): 62–65.

Stanton, E. 1895. The era of the bicycle. *The American Wheelman* 41 (30 May 1895).

Suffrage autoists motor 10,700 miles. *The New York Times* October 1, 1916.

Todd, H. 1911. Getting out the vote: an account of a week's automobile campaign by women suffragists. *American Magazine* 72 (Sept. 1911): 611–619.

Townsend, J. 1895. The social side of bicycling. *Scribner's Magazine* 17, 6: 704–708.

Trout, G. 1920. Side lights on Illinois suffrage history. *Journal of the Illinois State Historical Society* 13, 2: 145–179.

'Votes for Women' to be Discussed at Costa's Ballroom This Evening. *The Athens Daily Herald* 4, 205 (April 18, 1916): 1.

Ward, M. 1896. *Bicycling for Ladies: The Common Sense of Bicycling.* New York: Brentano's.

Waugh, A. 1900. The old road map. In *Legends of the Wheel*, pp. 107–110. Bristol, UK: J.W. Arrowsmith.

Weeks, C. 1906. The ABC of the automobile. *Outing* 47, 6: 687.

Wells, H.G. 1896. *The Wheels of Chance: A Holiday Adventure*. London: J.M. Dent/ New York: The Macmillan Co.

What some women are doing: suffrage and the war result in new feminine history— Halloween is not forgotten. *Motor Age* 30, 16 (Oct 19, 1916): 28.

Wheel whirls. *Godey's* 135, 805 (1897): 105.

Where women motorists excel mere men. *The Literary Digest* 4 December 1920, 73, 76 and 80.

Why women are, or are not, good chauffeuses. 1904. *Outing* 44, 1: 154–159.

Willard, F. 1895/1991. *How I Learned to Ride the Bicycle: Reflections of an Influential 19th Century Woman*. Ed. C. O'Hare. Sunnyvale, CA: Fair Oaks Publishing.

Wisby, H. 1905. Camping out with an automobile. *Outing* 45, 6: 739–745.

Wisconsin Tour and Hand Book. 1897. Compiled and arranged by S. Ryan. Appleton, WI: Wisconsin Division, League of American Wheelmen.

Woman drags road. *Southern Good Roads* September 1912, 28.

Woman road builders. 1912. *Twentieth Century Magazine* 5, 5 (March): 88.

The Woman's National Old Trail Roads Association. *Southern Good Roads* December 1911, 19.

Wood, M. 1913. What women's clubs are doing. *Ladies' Home Journal* 30, 2: 28.

Workman, F. 1892. Bicycle riding in Germany. *Outing* 21, 2: 110–111.

The world awheel. 1896. *Munsey's* 15, 2: 131–159.

Bibliography

Akerman, J. 1993. Blazing a well-worn path: cartographic commercialism, highway promotion, and automobile tourism in the United States, 1880–1930. *Cartographica* 30, 1: 10–20.

———. 2002. American promotional road mapping in the twentieth century. *Cartography and Geographic Information Science* 29, 3: 175–191.

Aronson, S. 1952. The sociology of the bicycle. *Social Forces* 30, 3: 305–312.

Bauchspies, W. and M. Puig de la Bellascasa. 2009. Feminist science and technology studies: A patchwork of moving subjectivities. An interview with Geoffrey Bowker, Sandra Harding, Anne Marie Mol, Susan Leigh Star and Banu Subramaniam. *Subjectivity* 28: 334–344.

Bauer, J. 2009. The *Official Automobile Blue Book*, 1901–1929: Precursor to the American Road Map. *Cartographic Perspectives* 62 (Winter): 4–27.

Bogardus, R. 1998. The reorientation of Paradise: modern mass media and narratives of desire in the making of American consumer culture. *American Literary History* 10, 3: 508–523.

Clarke, D. 2003. Women on wheels: 'a threat at yesterday's order of things.' *Arizona Quarterly* 59, 4: 103–133.

Clarsen, G. 2008. *Eat My Dust: Early Women Motorists*. Baltimore, MD: The John Hopkins University Press.

Cowan, R. 1983. *More Work for Mother: The Ironies of Household Technology from the Open Hearth to the Microwave*. New York: Basic Books, Inc., Publishers.

Cresswell, T. 2013. *Geographic Thought: A Critical Introduction.* Chichester, West Sussex, UK: Wiley-Blackwell.

Epperson, B. 2000. Failed colossus: strategic error at the Pope Manufacturing Company 1878–1900. *Technology and Culture* 41, 2: 300–320.

Finison, L. 2014. *Boston's Cycling Craze, 1880–1900: A Story of Race, Sport, and Society.* Amherst, MA: University of Massachusetts Press.

Foster, M. 1999. In the face of 'Jim Crow': prosperous Blacks and vacations, travel and outdoor leisure, 1890–1945. *The Journal of Negro History* 84, 2: 130–149.

Fry, A. 1969. Along the suffrage trail: from west to east for freedom now! *American West* 6, 1: 16–25.

Garvey, E. 1995. Reframing the bicycle: advertising-supported magazines and scorching women. *American Quarterly* 47, 1: 66–101.

Golledge, R. 1999. Human wayfinding and cognitive maps. In *Wayfinding Behavior: Cognitive Mapping and Other Spatial Processes*, ed. R. Golledge, pp. 5–45. Baltimore/London: The Johns Hopkins University Press.

Gudis, C. 2010. History and technology forum: driving consumption. *History and Technology* 26, 4: 369–378.

Kerber, L. 1988. Separate spheres, female worlds, woman's place: the rhetoric of women's history. *The Journal of American History* 75, 1: 9–39.

Koelle, A. 2010. Pedaling on the periphery: the African American Twenty-Fifth Infantry Bicycle Corps and the roads of American expansion. *Western Historical Quarterly* 41: 305–326.

Kolbert, E. 2008. Place settings. *New Yorker* 84, 336: 88–92.

Levenstein, H. 1998. *Seductive Journey: American Tourists in France from Jefferson to the Jazz Age.* Chicago/London: University of Chicago Press.

Mackintosh, P. and G. Norcliffe. 2007. Men, women and the bicycle: gender and social geography of cycling in the late nineteenth-century. In *Cycling and Society*, eds. D. Horton and P. Rosen, pp. 153–177. Farnham, UK: Ashgate.

Macy, S. 2011a. *Wheels of Change: How Women Rode the Bicycle to Freedom (With a Few Flat Tires Along the Way).* Washington, DC: The National Geographic Society.

———. 2011b. The devil's advance agent. *American History* 46, 4: 42–45.

Mayo, E. 1991. "Do everything" the life and work of Frances Willard. In *How I Learned to Ride the Bicycle: Reflections of an Influential 19th Century Woman*, ed. C. O'Hare, pp. 1–13. Sunnyvale, CA: Fair Oaks Publishing.

McConnell, C. 2000. *'A Reliable Car and a Woman Who Knows It': The First Coast-to-Coast Auto Trips by Women, 1899–1916.* Jefferson NC/London: McFarland & Company, Inc., Publishers.

McGaw, J. 1989. No passive victims, no separate spheres: a feminist perspective on technology's history. In *In Context: History and the History of Technology, Essays in Honor of Melvin Kranzberg*, eds. S. Cutcliffe and R. Post, pp. 172–203. Bethlehem, PA: Lehigh University Press.

Medlicott, C. and M. Heffernan. 2004. 'Autograph of a nation': the Daughters of the American Revolution and the National Old Trails Road, 1910–1927. *National Identities* 6, 3: 233–260.

Morin, K. 2008. *Frontiers of Femininity: A New Historical Geography of the Nineteenth-Century American West.* Syracuse, NY: Syracuse University Press.

Olliff, M. 2010. "The most famous Good Roads woman in the United States": Alma Rittenberry of Birmingham. *The Alabama Review* 63, 2: 83–109.

Petty, R. 1995. Peddling the bicycle in the 1890s: mass marketing shifts into high gear. *Journal of Macromarketing* 15, 1: 32–46.

———. 1997. Women and the wheel: the bicycle's impact on women. In *Cycle History: Proceedings of the 7th International Cycle History Conference*, ed. R. van der Plas, pp. 112–133. San Francisco, CA: Rob van der Plas.

Preston, H. 1991. *Dirt Roads to Dixie: Accessibility and Modernization in the South, 1885–1935*. Knoxville, TN: The University of Tennessee Press.

Pridemore, J. and J. Hurd. 1995. *The American Bicycle*. Osceola, WI: Motorbooks International Publishers & Wholesalers.

Ramsey, A. 2005. *Alice's Drive: Republishing Veil, Duster, and Tire Iron*. Annotated by G. Franzwa. Tuscon, AZ: The Patrice Press.

Ritchie, A. 2003. The League of American Wheelmen, Major Taylor, and the 'color question; in the United States in the 1890s. *Culture, Sport, Society* 6, 2–3: 13–43.

Rogers, T. 1972. The bicycle in American culture. *Contemporary Review* 221, 1281: 200–203.

Scharff, V. 1991. *Taking the Wheel: Women and the Coming of the Motor Age*. New York: The Free Press.

———. 1999. Lighting out for the territory: women, mobility and Western place. In *Power and Place in the North American West*, eds. R. White and J. Findlay, pp. 287–303. Seattle, WA: Center for the Study of the Pacific Northwest in association with the University of Washington Press.

———. 2003. *Twenty Thousand Roads: Women, Movement and the West*. Berkeley, CA: University of California Press.

Seiler, C. 2009. *Republic of Drivers: A Cultural History of Automobility in America*. Chicago, IL: University of Chicago Press.

Sewall, J. 2003. Sidewalks and store windows as political landscapes. In *Constructing Image, Identity and Place*, eds. A. Hoagland and K. Breisch, pp. 85–98. Knoxville, TN: University of Tennessee Press.

Smith, R. 1972. *A Social History of the Bicycle: Its Early Life and Times in America*. New York: American Heritage Press.

Spreng, R. 1995. The 1890s bicycling craze in the Red River Valley. *Minnesota History* 54, 6: 268–282.

Strange, L. and R. Brown. 2002. The bicycle, women's rights, and Elizabeth Cady Stanton. *Women's Studies* 31: 609–626.

Teape, V. 1980. The road to Denver. *The Palimpsest* 61, 1: 2–11.

Tobin, G. 1974. The bicycle boom of the 1890s: the development of private transportation and the birth of the modern tourist. *The Journal of Popular Culture* 7, 4: 838–849.

Watts, L. 2008. The art and craft of train travel. *Social & Cultural Geography* 9, 6: 711–726.

Wosk, J. 2001. *Women and the Machine: Representations from the Spinning Wheel to the Electronic Age*. Baltimore/London: The Johns Hopkins University Press.

3   "Woman's work" Part I – American women missionary geographies

The phrase "woman's work" has many implications. In the Progressive Era, it could mean the work women conducted in the home (cooking, cleaning, laundry, meeting the needs of children), but it also could refer to work women conducted outside the home, work that women were viewed as uniquely qualified to carry out (Figure 3.1). This chapter and the next will examine "woman's work," the spaces in which this work took place, and the geographic knowledge created in these contexts. In particular, I will be examining two areas described as woman's work: missionary outreach and settlement work.

The journal *Woman's Work* (formerly *Woman's Work for Women*) was published by the Woman's Foreign Mission Society of the American Presbyterian Church beginning in 1870, choosing as their title "the phrase that both overseas missionaries and their supporters in North America used most often to describe women's missionary activity" (Flemming 1989, 1). Settlement work, such as practiced by the Residents of Hull-House, involved not only providing educational and socializing opportunities for neighborhood citizens but also carrying out neighborhood studies to determine the needs of the community. These studies involving door-to-door surveys were viewed as "woman's work" by professional sociologists (Deegan 1988, 46). Both practices of woman's work took place in sites where knowledge was being actively constructed. In the course of their work, women created and actively circulated geographic knowledge at a variety of scales.

As discussed in Chapter 1, women with education had few employment choices leading up to the Progressive Era, teacher and missionary being the classic examples. In the early to mid-1800s, women who chose the missionary route were most often expected to marry and to play a supportive role to the missionary husband. By the late 1800s and the beginning of the Progressive Era, more and more women were becoming missionaries, to the point that by 1900 women dominated the missionary field, both as in-the-field missionaries and as members of missionary organizations. As part of their efforts to support and advance missionary work, women missionaries published accounts of their work in a variety of forums, as articles in missionary magazines such as *Woman's Work* and later in life in book form. These accounts often

THE WEAKER SEX

Figure 3.1 "The Weaker Sex" illustration first appeared in a woman's suffrage issue of *Puck*, an American humor and satire magazine (February 1915). It later was used on the cover of the Women's Christian Temperance Union magazine *The Union Signal* (30 September 1915).

incorporated geographic information in their descriptions and sometimes were illustrated with maps. In American churches, women's missionary groups engaged in a variety of activities in support of their "home" and "foreign" missionaries, with geographic knowledge and maps central to their work.

Jane Addams, looking for an alternative to teaching or missionary work, popularized the settlement house movement, establishing Hull-House in 1889 with her friend Ellen Gates Starr. Addams and Starr were looking for meaningful work for themselves while significantly contributing to society. Inspired by Toynbee Hall in London, they began working on Chicago's South Side. As Hull-House Residents worked to improve their community, they carried out studies of the neighborhood, often mapping the results, creating and circulating geographic knowledge as part of their work. Addams and Hull-House led to the development of a "settlement house movement": Progressive-minded men and women across the country, and even beyond, established "settlement houses" modeled on Hull-House, studied their communities and worked in their neighborhoods to improve the quality of life (Ogawa 2004).

The transformations of U.S. cities, brought about by the forces of capitalism, urbanization, industrialization, and immigration, created new social problems, such as poor living conditions in tenements, unregulated labor especially of women and children, and growing problems of poverty. The problems Progressive era women and men saw in their country prompted many to find some means of alleviating and improving these situations. For some social reformers, it was an opportunity to combine their Christian theological beliefs with the desire to address these problems (Lindley 2006, 1069). They saw social structures as impeding individuals' progress toward Jesus Christ. By addressing these social problems using both Christian beliefs and modern social science, they might "establish the kingdom of God in this world, rather than the next," in what was termed the "social gospel" movement (Rynbrandt 1998, 73).

Both missionary work and settlement work can be seen as part of the social gospel movement, sometimes called "social Christianity, social evangelism, or applied Christianity" (Williams and Maclean 2012, 342). Social gospel is the combination of theology and social action, based on the theological belief that the gospel taught by Jesus emphasized not just the salvation of individual sinners but the:

> ... redemption of sinful social systems and structures, especially economic and political ones ... believers in the Social gospel message analyzed the ills they perceived in American society and devised reform strategies that would bring their vision into the Kingdom of God— one that manifested social justice, economic equality, and political freedom—into reality.
>
> Gifford 2006, 1027

Both men and women activists of the Progressive Era believed in the social gospel and worked to bring about "social salvation." But women who followed the social gospel often believed women had a "special—often morally superior—nature and women's special responsibility for the family" which made them especially well-equipped for this work (Lindley 2006, 1075).

While often evoked as a singularity, the social gospel was anything but, arising from eighteenth century shifts in thinking about human rights, religion, labor, and humanitarianism. Its major tenets can be described as:

> ... (1) belief in the innate goodness of humankind; (2) acceptance of evolution as compatible with God's plan for the universe; (3) rejection of the determinism of evolution in favor of the idea of development of

progress; (4) belief in the inevitable progress of society; (5) redefinition of the Kingdom of God as an earthly utopia; and (6) belief that the Kingdom would be established in the United States.

Williams and Maclean 2012, 344

But men and women differed widely in their approaches to the social gospel with "… women presumed responsibility for social conditions that affected women, children, and family life. Men dutifully maintained oversight of industrial, political, and theological concerns" (Edwards and Gifford 2003, 5). As women's colleges produced educated women, it became a balancing act justifying educating women at a time when traditional roles defined separate spheres for men and women.

Service work, especially combined with a missionary aspect and focusing on women and children, was viewed as non-threatening and socially acceptable for women, offering an outlet for educated women (Williams and Maclean 2012, 353). Many women followers of the social gospel focused more pragmatically on action, rather than on theology, studying communities to determine problems and potential solutions, giving speeches, writing articles, publishing pamphlets and books (Lindley 2006, 1071). They educated themselves on how to conduct surveys, how to display their data as tables, charts, and maps, and how to use this evidence to argue persuasively for change. They worked with church organizations, home missions and supported foreign missions. They joined women's clubs and some joined organizations focused on particular reforms, such as the Woman's Christian Temperance Union, the American Woman Suffrage Organization, and the National Woman's Trade Union League (Gifford 2006, 1028). More dedicated individuals might become a missionary or work at a social settlement, working for a few months or years, for others, finding lifelong careers (Lindley 2006, 1072). But not all members of organizations or workers at settlement houses were social gospel supports, just that such organizations and work offered women an opportunity to practice the Social gospel (Lindley 2006, 1075). In fact, women who practiced the social gospel varied widely in their beliefs, their practices, their backgrounds, their agendas … in almost everything (Edwards and Gifford 2003, 5).

Through the woman's work carried out by missionaries and settlement workers, similar yet very different practices, geographic knowledge was created and circulated by Progressive Era women. I will begin by discussing the geographic work of women missionaries. I will consider settlement house contributions in Chapter 4.

"… needed everywhere": Missionaries and geographic knowledge

The world is open to Christian women as it never has been before. She can go almost anywhere, and she can engage in almost every kind of

work. She is needed everywhere. She must write; for a literature must be
created for the women of the East. She must teach; for the convert must
be trained, and the heathen won. She must evangelize; for her feet alone
can carry the good tidings of peace to her sisters in their seclusion... .
Bishop Thorburn, quoted in Lilly Rider Gracey's *Gist: A Hand-book of
Missionary Information* (1893)

Integral to the practice of Christianity, even from its earliest days, was the
mandate to spread the word of Christianity, to share the "good news." As Europeans, and eventually Americans, grew in wealth and power and exerted
their power over other less developed countries through colonialism and imperialism, missionary work became integral to the process. As they spread
the word of Christianity, saving the "unenlightened" in developing parts of
the world, they would simultaneously uplift these populations through Western ideas of modernity. Missionary work among Europeans and Americans
went hand-in-hand with imperial expansion with missionaries serving as
"mediators" between colonial powers and local populations (Endfield 2011,
204). Through missionary societies affiliated with large denominations (such
as Congregationalists, Methodists, Baptists, Lutherans, Episcopalians, African Methodists, and Presbyterians), missionaries were supported and sent
out to the developing world (Robert 2006, 837). Missionaries were often
among the first to enter "frontier" regions and became valuable sources on
the region's physical and cultural geography. A 1905 poem on "The Good
Old Native Preacher" describes "him" as "... a map and a directory, a walking gazetteer" (Gilmore 1905). This information was distributed through the
missionary's publications (letters, diaries, articles, books) created to garner
support for their endeavors. Missionaries often created maps and charts
connected with their missionary work, from plans of mission compounds
to the routes taken on mission trips into interior regions. These maps might
be submitted with reports to their supervisors or published in magazines
or books. The definitive geographer/missionary was David Livingstone, exploring central Africa and advancing British knowledge of its empire while
spreading the word of Christianity. Jesuit Matteo Ricci was another geog-
rapher/explorer recognized as a scientist and cartographer while advancing the Christian mission in China (Livingstone 2005, 53 and 55). A select
few missionaries, like Livingstone, produced some of the very first maps of
regions, in work essentially disconnected from their mission work (Braun
2012, 250). Particularly in the mid-nineteenth century, there was significant
overlap in the interests of missionaries, British officials, and businessmen in
advancing "civilization" while at the same time expanding the geographic
knowledge of the age, resulting in the interactions between such workers and
the Royal Geographical Society (Bridges 2008). The scientific contributions
of male missionaries has been explored by geographers such as David N.
Livingstone, Felix Driver, Clive Barnett, and Georgina Endfield (Barnett
1998; Driver 1999; Endfield 2011; Livingstone 2005).[1]

Missionary women were more often involved in what might be described as mundane acts of geographic knowledge creation and circulation, writing largely for audiences of other women. Early in the nineteenth century, women interested in missionary work were restricted to the role of wives of ordained ministers or male missionaries. Women's education in the United States grew over the course of the nineteenth century but the opportunities to use their educations remained circumscribed, with education largely focused on preparing future teachers, mothers, and missionaries. Mount Holyoke, a pioneering women's seminary in South Hadley, Massachusetts, had from its origins in 1837 an objective of cultivating "the missionary spirit among its pupils" (Hill 1985, 42). It is estimated between 1839 and the 1950s, Mt. Holyoke had over 500 alumnae engaged in missionary work.[2] Women's roles in many Protestant denominations were limited yet women yearned for meaningful activity and searched for active opportunities (Garrett 1982, 222). Beginning in 1861, women organized "separate women's mission boards to support their own global agendas to evangelize and education women and children around the world" (Reeves-Ellington 2011, 192). In particular, American women were concerned women in non-Christian countries were "degraded and that only through the support of their Christian sisters could they enjoy the same status that women in Christian countries presumably enjoyed" (Cramer 2004, 124). They asserted, as women, they were best suited to reach the women in other countries (Cramer 2003, 213).

I am focusing on Protestant women missionaries and the widespread mission movement mobilizing American women beginning in the 1860s. The experiences of Catholic women were different: "The predominant way for an American Catholic woman to become a missionary or evangelist was to enter a religious order... ."(Dries 2006, 843). For Protestant women, there was a number of avenues for them to become a missionary: training institutes and colleges, training to be medical missionaries, opportunities for missionary work in the United States, and foreign missionary work, and even just supporting missionary work through fundraising and mission study at their home churches, none of which required them to join an order. American Protestant women were drawn to these opportunities for service. It is estimated by 1900, there were 6,000 missionaries from the United States internationally, of which 2/3rds were women (Reeves-Ellington 2011, 191; Robert 2006, 837).

Many women's mission organizations actively recruited. Pamphlets, books, and textbooks were created to encourage women to consider missionary work, such as Lilly Ryder Gracey's *Gist: A Hand-book of Missionary Information* from 1893, designed to be used in Bible study groups while promoting women's missionary work. An article in the 1897 *Woman's Missionary Friend*, details missionary training schools across the country, including the Folts Mission Institute in New York, the Bible Normal College in Springfield Massachusetts, the Chicago Training school, and the Scarritt Bible and Training School of Kansas City (North 1897).[3] At such institutions, students were trained in Bible Study, prayer, and practical work, with some offering coursework in such topics as church history, "applied Christianity," pedagogy, practical medicine, and bookkeeping (North 1897). Biblical geography and "map talks" were part of the curriculum. At Ingleside Seminary, a black boarding school, a teacher commented "We always have maps and map-talks at every meeting. The instruction along these lines will be of great use to girls when they go out and become church workers" (*Home Mission Monthly*, April 1900).

Missionary work also offered interested American women the opportunity to enter the field of medicine, at a time when women physicians could not get jobs in the United States (Robert 2006, 837). The American Medical Missionary Society published the pamphlet *A Great Field for Women* encouraging women to apply to their medical program which would train them to be doctors and send them out as Medical Missionaries (188-). Thus the need for women missionary-physicians led to the expansion of medical training for women in the United States (Robert 2002, 70).

Whatever the training, be it basic missionary or medical training, women missionaries could expect additional location-specific training once they reached their mission. First and foremost would be learning the language, important to both negotiating the country as well as reaching their parishioners: "Before the impatient missionary could think of beginning her school she must be a pupil" (Montgomery 1910, 89). In some locations, such as Papua New Guinea, some women "reported preaching in faltering Dobuan within weeks of their arrival" (Langmore 1982, 143). More complex languages, such as Japanese and Mandarin Chinese required much longer time to master. Reviewing records of Congregational missionaries in Japan, scholar Sandra Taylor found "… missionaries frequently commented that at least three years of study in Japan was necessary to gain sufficient language ability to work well with the people" (Taylor 1979, 34). Missionaries would stay usually at another mission as they learned the language, studying with a private tutor (Flynt and Berkley 1997, 61). In a few months, they might be functional on an elementary level but it would take much longer to master the language at a level to convey complex ideas such as those found in religion. As they learned the language, they would become familiar with the culture and their customs, assisting established missionaries in a sort of internship before they would be sent out to their main mission location and began the real work (Flynt and Berkley 1997, 56 and 58). Language skills were essential to succeeding at their tasks. Mary Wainwright in her article on her work in Okayama mentions: "From morning till night, unless I am with one of the foreigners, I rarely use my mother tongue" (Wainwright 1902, 295).

At the missions, the women would engage in a variety of work. While teaching and evangelizing were central to their practice, they also might be called upon to manage budgets, oversee construction of facilities, provide basic medical care, and serve as models of what it meant to be a "modern" woman as well as a Christian woman. Their evangelizing often involved extensive public speaking and traveling considerable distances into "frontier" zones but also on a more intimate scale, home visits, speaking and teaching women in either women's spaces in the home, or in women-designated spaces in the mission.

The term "missionaries" is more often associated with overseas work but "home missions," missions within the United States, were an important aspect of most American women's mission boards. In particular, women missionaries were sent out to work: on Native American reservations; with Black Americans in the South; with White Americans in Appalachia; amongst recent immigrants to the United States; and amongst the Mormons in Utah. While both Black Americans and those in Appalachia were Christians (and, theoretically, Mormons also), their practices were seen as "deviant forms of Christianity. Black Americans and Appalachians heard missionaries describe their churches as superstitious, primitive, and overly emotional" (Fraley 2011, 30). In Appalachia alone, it is estimated by 1920, at least seventeen denominations had missions with over five hundred missionaries. It was not uncommon for a woman to begin with a home mission before taking on a foreign mission, such as Nellie Arnott who first had a mission in Georgia before being sent to work in Angola (Hill 1985, 126; Robbins and Ellis Pullen 2011).

Overseas, American women missionaries were sent to work in China, Japan, Korea, Siam, India, and locations in the Middle East, South America, and Africa. Many of the women missionaries opened schools and taught classes for women and children, teaching subjects such as reading, writing, arithmetic, geography and domestic skills such as sewing (Langmore 1982, 143). They established clinics and hospitals. Eventually they established high schools, seminaries or theological schools, and vocational training schools (Cox 2008, 87). They trained and worked with native "Bible women" who extended the reach of their efforts (Taylor 1979, 32).[4] Women missionary doctors not only saw patients but often established hospitals and provided training in Western medicine, often founding the first women's medical institutions in Asia (Flemming 1989, 4).

In many Asian cultures, women were spatially segregated and had been largely unreached by male missionaries. In India, "zenana" referred to spaces in upper-middle-class Hindu and Muslim family homes "… marked off exclusively for women, where they would cook, do domestic work, and spend leisure hours" (Singh 2005, 128). Women missionaries literally could go where no (Western) man had gone before, entering spaces forbidden to males, and reaching audiences viewed as critical for the success for missionary work.

Some women missionaries played more active roles in advancing the church. Helen Emily Springer walked 1,500 miles across central Africa in 1907 with her husband, "… scouting out places for the future Congo mission of the Methodist church" (Robert 2002, 67). Women medical professions not only practiced medicine but also played important administration and organizational roles. In India, Rose Greenfield, Clara Swain and Anna Kugler

> ... worked on the expansion, architectural design, and planning of their hospitals. They all were sensitive to the specific cultural and climatic requirements for a hospital in South Asian context. Innovative in the planning and organization of their hospitals, they ingeniously adapted local materials and resources for their purposes.
>
> Singh 2005, 132

Through their work to locate new missions and see to the construction of new spaces of missions, clinics, and hospitals, they were literally as well as figuratively building the church.

Periodically, missionaries would return home to see family and rejuvenate. On these home trips, they would be sent out on speaking tours, addressing church services, conferences, camp meetings, Sunday schools and women's study groups (Robert 2002, 67). By appearing "in person," they were verifying their printed narratives, reinforcing them with their oral testimonies and sometimes visually too, through their photographs and magic lantern shows. Through their presentations, they were educating their audiences on geography and life overseas and on what women can do. A sense of this is conveyed through the mission periodicals, which in addition to reports written by missionaries, included descriptions of state and city chapter activities, often listing the visiting missionary presentations. On a much larger scale were "missionary expositions," such as the 1900 Ecumenical Conference on Foreign Missions (ECFM) in New York and the 1911 "World in Boston," where thousands of missionaries gathered to offer "vivid testimonials" to thousands of Christians.[5] The 1900 ECFM drew an estimated 200,000 Protestants, who came to hear the testimonials and take in the "missionary exhibits" from around the globe (Hasinoff 2010, 84–5).

Some missionaries, however, never returned home. Exotic diseases, overwork, and sometimes violence led to their deaths in the field. In China in 1899–1901, unrest led to violence against Christian missionaries: an estimated 250 Christian missionaries, including Catholic priests and nuns and Protestants missionaries and their families, were killed (estimates of the Chinese Christians killed at the time are over 30,000) (Klein 2013). In the missionary press, those killed were viewed as "martyrs" – individuals who were killed for their beliefs or for refusing to renounce their beliefs (for example, see Bashford 1905b). Despite the dangers, many women found the work rewarding and dedicated their lives to their work, returning again and again to their missions.

For numerous women, mission work not only provided them an education and a chance to apply it, it also permitted travel, a degree of independence, and even opportunities for leadership, leading prayers and giving sermons (Robert 2006, 837). The mission work allowed them to use

> … religion as a platform to enter the more public sphere of missionary work … religion may have served as a Trojan Horse, allowing them to enter a social and political realm disguised in a religious cloak of propriety, service, and influence over others.
>
> Cramer 2004, 126

For a few, their mission work led to their ordination, as some non-Western churches ordained women long before women could be ordained in the United States (Robert 2002, 71).

Missionary woman and their practices of geographic knowledge creation

In October 1902, American missionary Mary Wainwright took her readers on a virtual tour of Okayama, in an article in *Mission Studies, Women's Work in Foreign Lands*:

> Leaving the park we cross the Asahi river which divides the city. Most of the year it is a beautiful stream, but once or twice a year it rises to such an extent that people watch it anxiously, wondering what damage it is going to do … We soon reach the preaching place on the corner. The preaching place is also my home, for I am now living over the church or preaching place… . There is not time to stop longer here, so we will go on past the Government Hospital and large Medical School, past the Government High school … If we wish to see anything of the surrounding country we will have to stop here, and call our jinrikishas and start for Saidaiji. It is a strong Buddhist center but there are a few faithful Christian there, and we wish to help them 'hold up their hands.' It is a pretty ride, and especially so at this time of the year, with the fields of yellow rape, the purple alfalfa and the different greens of the wheat and vegetables. The map of the province will also show you Kagato a little beyond Saidaiji… .

Wainwright's virtual tour of Okayama and beyond demonstrates how women missionaries shaped the geographical imaginations of their readers, American women, constructing vivid images of distant lands and even providing some sense of spatial relationships. Communication was an essential skill of missionaries. Not only did they have to establish a rapport with their target populations and their congregations, they also had a wide range of writing tasks, which might include learning the local language and translating the Bible and other texts into it, writing reports to mission organizations, publishing articles for mission periodicals, and often keeping a personal diary that might be someday turned into a published memoir.

Publications were an important aspect of the missionary work that most missionaries engaged in, a means of reaching a much broader audience, both in their mission country and back home. Missionary publications had a wide range of aims and functions, from reinforcing a shared religious identity and disseminating religious propaganda, to circulating information about peoples and landscapes, shaping readers' geographical imaginations about the world and about the missionary work with these peoples and landscapes (Jensz 2012, 234–5). Missionary boards produced a variety of materials, from newsletters/magazines to books and teaching materials. Missionaries initially wrote letters back to home congregations and friends that would be circulated as an informal source of information about missionary work and the experiences individuals were having in the course of their work. By the late nineteenth century, printing had become inexpensive and technology now allowed the mass printing of photographs. Various women's church groups as well as individuals would subscribe to these magazines.

Missionary newsletter/magazines, such as *Woman's Work*, were filled with articles about missionary activities, such as descriptions of the cultures, landscapes, religious practices, as well as accounts from the missionaries about their work, and testimonials from new converts. In addition, there were short reports from other missions, accounts of meetings held, descriptions of state activities, and often teaching materials. These teaching materials were often for a variety of levels, from women's meetings to "young people" and children with the lessons employing articles from the magazine as the basis of discussion. Maps and discussion of the geography of the locations was a central aspect of the lessons. These lessons, whether for adults or for children, were designed to be educational, to promote mission work (encourage donations and attract readers), and to reinforce their religions' position on various topics, including the role of women in society. Periodicals, with their immutable form, crafted by religious publishing houses, and distributed across the country, provided "… the best medium for disseminating information of the state and needs of the mission field to a broad range of people over geographically dispersed areas" (Jensz 2012, 234–5 and 239). In their writings, women missionaries had to convey a complicated message, encompassing a compelling message of need, creating a sense of difference, and often crafting a sense of place.

These materials could be read in the home and/or used as part of women's missionary meetings, where they would study information from the publications and discuss it with other members. Dana Robert found in her research on a First Congregational Church of Manchester New Hampshire that

> different women in the group became recognized local experts in different parts of the world and presented regular programs on Christian issues in those places. Regular correspondence with particular missionaries were features in the programs, and women prided themselves on their knowledge of the church in other lands.
>
> Robert 2002, 75–6

On a larger scale, summer mission schools were held at various locations across the country, attracting thousands of women who "studied the mission text and enjoyed fellowship, pageants, and prayer" (Robert 2006, 839).

Returning to the October 1902 issue of *Mission Studies* containing Miss Wainwright's virtual tour of Okayama, the issue also included an overview of Wainwright's work in Okayama, a description of her home/mission space, an article explaining the basics of Buddhism and Shintoism, an article discussing a summer vacation in the mountains for a mission teacher at Kobe College, and a reference map (Figure 3.2). The Young People's and Children's Lesson continues the Japanese theme and provide more information about the mission activities and the geography. Viewed together, an overview of the physical and cultural geography of Japanese regions is created. Much of the information they presented is general geographic knowledge, basic facts about the country, the region, and the town and its landscape and culture. While crafting a sense of place and contributing to the geographical imaginations of their readers, they were also building connections between the missionaries in Japan and their readers in the United States. Wainwright's use of "we" as she takes her readers on a virtual tour of Okayama and refers them to the map creates a sense of intimacy.

Similarly, the article by the mission teacher begins "I wish in reply to your kind letter I could send you a bit of the beauty that is surrounding me on all sides" (Chandler 1902, 301). The casual, chatty tone, the references to "we" and "your kind letter" creates a sense of friendly community. But there are also references woven in to the American landscape that the American audience would understand, such as "Afraid? No, indeed, we felt as safe and secure as though we were in the White Mountains, in staid New England" (Welpton 1902, 304). And while discussing their work overseas, there is the acknowledgement of the support of the American women: "It is a comfort to know you all are bearing this work on your hearts and prayers, that it is yours as well as ours… ." (Parmalee 1902, 312).

To further explore the geographies constructed in women's missionary publications, I conducted a brief comparison study of men's and women's articles about China, using a limited sample of eleven articles from women's missionary publications and eleven articles from general missionary publications by male authors, published between 1890 and 1925, and written for American audiences.[6] Effort was made to equally represent the time period, as well as capturing a range of Protestant denominations, and a variety of publication forms (magazine articles and pamphlets).

In general, both women and men presented in their publications to some degree the regional geography of China. There is a sense of the site and situation, including sometimes discussions of the major features, waterways, and available resources. Nearly all address the size of the Chinese population. There is some discussion of language, especially regarding the development of a phonetic alphabet to facilitate communication across

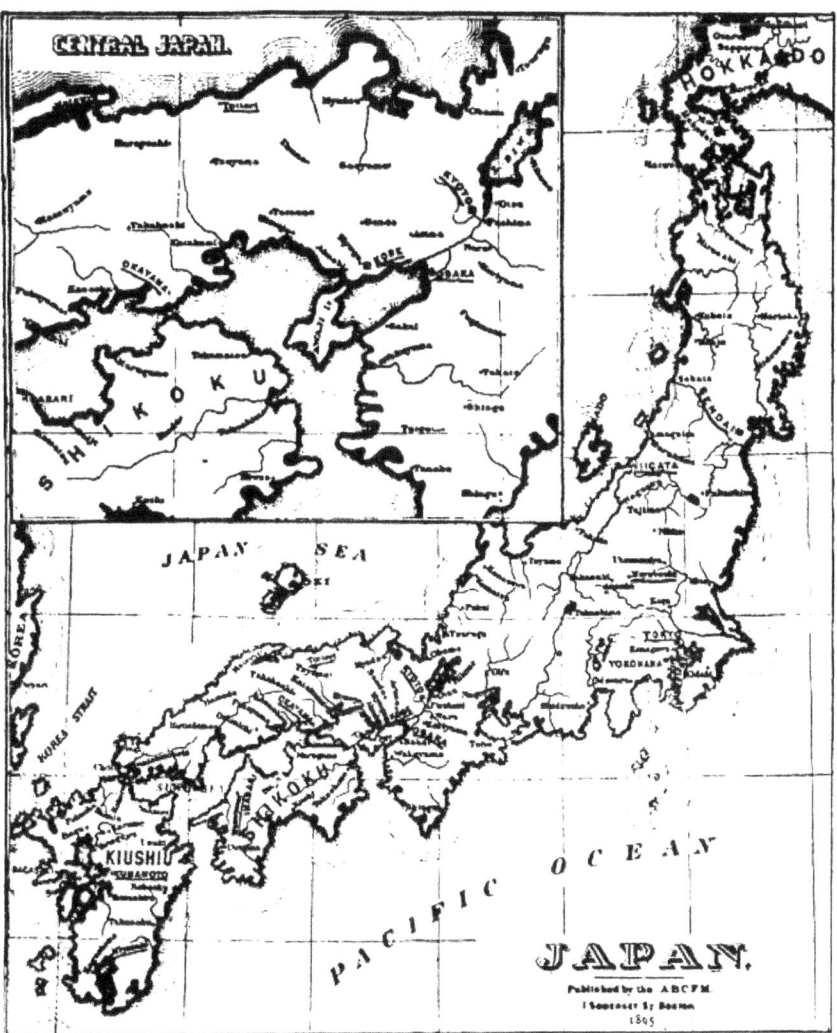

Figure 3.2 Map of Japan from the October 1902 issue of *Missions Studies, Women's Work in Foreign Lands* that features Mary Wainwright's articles on her work in Okayama.

China. All address to some extent, some more than others, the religions traditionally practiced – Animism, Confucianism, Taoism, Buddhism, "Muhammedism" – as well as general "idol worship." There was also discussion of certain Chinese cultural practices viewed negatively by Westerners, such as opium addiction, gambling, and footbinding. Both men and women included maps in their articles at times.

But there were gender differences when it came to discussing Chinese geography. Women writers tended to include more landscape descriptions, with some crafting very vivid senses of place. For example:

> As our little procession of blue-clad, trousered maids emerges from the compound gate, carrying their hymn books and Bibles tied in hand-kerchiefs of varied colors, the view that bursts upon us is well worth going some distance to see. All about us, rolling back in their majesty, are grand old rocky hills and mountains, covered here and there with huge masses of stubborn rock, and brightened now by a clump of crim-son azaleas, now by lilacs, and again by delicate lavender azaleas. Here and there are beautiful big wild roses, often forming hedges along the terraced fields.
>
> Down over the hillside "Little Angel" leads our buffalo cow from among the many that are grazing along the hillsides, and hangs his head in shy delight as he sees the Su-gus, who he is soon to follow to Sunday school. Birds of many odd and beautiful kinds are flying near-by, unafraid of either native or foreigner. One wings its way out over the plain and draws our eyes to the near-by Ohio-like view of level land immediately before us, ending in the un-Ohio-like Diong-loh pagoda far down in the distance.
>
> Dornblaser 1911, 370

Only the women writers mentioned landscape terracing, a stereotypical human-modification associated with the Chinese landscape. Women were also the only authors to mention China's climate, although in one instance they quoted another source describing China's climates as being similar to that of the United States (Wright 1897, 79) (note, too, the likening of the landscape to Ohio in the above quote). In another instance, it was to com-plain about the hot humid summers in Shanghai and affiliated disease out-breaks, necessitating the missionaries' retreat to the mountains for a few weeks (Rugh 1918, 119–120).

Women also captured the cultural geography of China relevant to women, such as foot binding, the unbinding of feet, infanticide, forced mar-riage, tolerance of wife beating, and the value of sons.[7] When the horrors of foot binding were discussed, emphasis was placed on the advances made to end the practice, including eventually the government prohibition of the practice (Bashford 1905a, 243). Photographs in women's writings included before and after images of bound feet, and a group photograph of Chinese women and missionaries featured bound-feet women in the front row, their feet clearly visible (Dornblaser 1911, 368; Snodgrass 1905, 28). By compari-son, foot binding was only briefly mentioned in the male author's writings. Critical accounts of Chinese parenting and hygiene, both associated with the "woman's sphere," were also discussed by women authors. Mentions were made of children smoking and piercing infant's ears, both frowned upon in mainstream American culture of the age (Benn 1912, 6; Dornblaser 1911, 371). There also was a great deal of distain of the hygiene, of the dirtiness of Chinese cities, of Chinese homes, of the unsanitary foods found in Chinese markets (Abbey 1905, 37; Dornblasher 1911, 366).

Writing for an audience of American women, women writers were addressing topics they knew would capture their audience's attention and possibly promote them to assist the missionaries in elevating the lives of Chinese women. Above all, they emphasized the mistreatment of Chinese women at the hands of their own culture, the virtuous and hard-working nature of Chinese women, and how, by helping Chinese women, they can elevate the whole of China – "Women educated will make China Christian" (Ogburn 1906, 165). One pamphlet, telling the story of "Ping-Kua, a girl of Cathay," explains girls are called "Ya-tou" translated as "female slave," capturing the subjugated role of women in traditional Chinese society (Benn 1912, 6).

Despite their emphasis of the mistreatment of Chinese women, the missionaries highlighted the advances made since their work began in China. Writing about the Tengchow mission, Mary Bergen comments on the progress made, with now second and third generation Chinese Christians coming to meetings at the mission: "we have the real cream at meetings: the pretty young wives and children of our teachers in college and schools... . You get intelligent answers from these women; their opinions are worth having" ("Missions and High Life ..." 1905, 28). There is the sentiment that it is an honor to be a missionary: "Oh! does it not seems strange that God, who loves them as He does us, still leaves us our privilege of being messengers to them when we have been so slow about it... .(Dornblaser 1911, 367). One author described missions as "A Social Power," that their impact was far beyond "individual lives but touches society as a whole, giving birth to new institutions and permeating non-Christian lands with a nobler spirit" (Fensham 1908a, 7). Yet, at least one women writer opined they could not just import Western culture and education, they had to develop what China and Chinese women need: "... a College that will meet the needs of the Chinese women and be not simply a replica of American or European institutions, is not built in a day" (The Woman's College ... 1920, 178).

Women authors discussed more their efforts working with local Christians, "Bible women," Chinese women converts to Christianity who assisted the women missionaries in their outreach work. In the majority of instances, these Bible women, whose efforts are praised and who seem to have been essential to the world, are unnamed, but there are exceptions. In a lovely tribute to a young Chinese woman described both as a "Bible teacher" and a "'new woman' of China," Emma Fleming, a missionary doctor, describes how "Beautiful arches dedicated to the memory of virtuous widows are one of the most common landmarks throughout old China" and how they now need "A New Memorial Arch" in the memory of the young women who "devote untrammeled lives to the service of their Master in various lines of work." Fleming goes on to tell the life story of Wang Su Djeng, who taught at the Louise Comegys Bible Institute at Ichowfu, Shangtung. Su Djeng was valued as a teacher and example, "who as triumphed through faith, rising from a weak girl to be the trusted Christian leader, the friend and counsellor of ministers and teachers, and one with whom we as 'sent ones' have delighted in sweet fellowship" (Fleming 1920, 171). Su Djeng succumbed to influenza and pneumonia in 1919, thus the need for "A New Memorial Arch."

Women had a stronger focus on education at all levels, at home and aboard. They provided lessons in the publications for American women, young people, and children to use in church meetings and Sunday schools. For example, as part of the September 1911 "Children's Department" section of *Lutheran Woman's Work*, the author "K.B.S." addresses the children and gives them directions: "First look up the history of the Great Wall.... . Then turn to Peking, the capital, where Emperor Hsuantung lives. From Peking you can take the railway, more than a thousand miles on a bee line, to Hankow, the great city in the heart of China, which Miss Peterson writes about... . (K.B.S. 1911, 373). K.B.S. continues, weaving together references to Chinese geography students could find on a map as well as references to articles appearing in this issue of the magazine. K.B.S. concludes:

> When you are through, if you have faithfully looked into your geographies and histories, you will be like the editor making up copy for this number of the magazine, you will know more about China than you ever dreamed of, and the great, old country will have a place in your hearts and prayers which it has never had before.
>
> K.B.S. 1911, 374

By comparison, only one or two of the general missionary magazines provided any lesson plans and these pale in comparison to the materials and directions provided in the women's magazines.

Women writers also addressed extensively their educational efforts in China. Geography is specifically mentioned as a topic they taught in their overseas schools. An advertisement in *The Chinese Recorder and Missionary Journal* details the geography teaching materials published by The Educational Association of China: *Mrs. Parker's Natural Elementary Geography, Mrs. Parker's Map-Drawing, Muirhead's Political Geography, Grave's Sacred Geography*, and *Maps of Scripture Geography*.[8] Mrs. Alice Parker's books were published by the Presbyterian Mission Press of Shanghai. One account of "News of the Year from Girls Schools," details the Tengchow High School was given "magnificent gifts" including "a map of China" and "a set of roller maps, in case" (News of the Year ... 1905, 35). Through their articles as well as their work, American women missionaries in China were teaching geography both in China and to audiences back home.

Male authors in their discussion of Chinese geography discussed resources more specifically, discussed the development of infrastructure and of modernization (such as advances in railways, telegraph and telephone).

They also discussed more the religions of China. Overall, male authors focused more on the "big picture," on geopolitics and their implications, on business/trade potential, and on modernization. They often used the language of conquest, tying it into their discussion of geopolitics or business, likening the missionary enterprise to a "military enterprise" (A Survey ... 1906, 9) and, it is not just China at stake, it is the whole of Asia: to command the situation in the Orient "we must take China" and "Why should the Church of Christ not be as aggressive in seizing strategic positions as the mikado's army?" (Our Four Fields 1906, 16). And there are calls for Christians to "invest" in China:

> THE GREAT INVESTMENT. How American capital is seeking the East! Exploring is followed by exploiting. Compared with commercial investment our fifty per cent increase would indeed be small. Compared with the millions spent in war or in building the Panama Canal, our investment seems insignificant. But when these loving gifts of the thousands of Christians are planted with prayer, God will bless, that they may grow and bear quicker, surer, larger harvests than in the business world.
>
> 'Open Door' 1906, 3

Writing in 1905, H.J. Openshaw suggests "While this large and productive prefecture is being exploited in the interests of modern commerce, we urge its claims as a field of missionary enterprise" (Openshaw 1905, 306). This focus on politics and business can be seen as attempting a more "masculine" take on missionary work: "From the perspective of other colonizers, who tended to cast imperialism as a manly act, the missionary enterprise was gendered as 'feminine'... . No matter how 'muscular' missionaries attempted to be, they were feminized by their ambiguous local alliances and domestic ideas" (Huber and Lutkehaus 1999, 12).

Ironically, in documenting the progress of missions statistically in terms of number of workers, baptisms, and income, the male writers capture the dominance of women in the missionary field at the time. While not addressing it in their text, their statistical tables clearly show more female missionaries than male. In one report, of the 828 people working with the mission, 497 are female to only 331 male (The Annual Report 1905, 75). Yet this remarkable fact is not commented upon. Eventually, the overwhelming presence of women in the mission field led to a movement to increase men's involvement in churches and missions, called the Men and Religion Forward Movement, a backlash against the "feminization" of American churches. Their slogan was "More Men for Religion, More Religion for Men" (Bederman 1989).

Returning to consider missionary writings and publications in general, the missionary authors believed their work was crucial to maintaining the financial as well as spiritual support network needed to make missionary work a success (Robbins and Ellis Pullen 2011, 69). With the importance of their writings in mind, they crafted their narratives for their multiple purposes and in multiple forms. In the case of Nellie Arnott, an American missionary in Angola:

> Arnott took advantage of the popularity and norms of women's travel writing. She also tailored features of travel writing for her movement's institutional agenda as missionary authors had been doing for decades. Accordingly, she wrote about several types of journeys simultaneously: literal travels to the location of her service (like her initial passage to Africa or her frequent treks to outstations), journey-like stages in her learning to be a successful foreign missionary, and the passage to Christianized, Americanized culture that she envisioned for the Umbundu people.
>
> Robbins and Ellis Pullen 2011, 72

Robbins and Ellis Pullen, in their study of the writings of Arnott, were able to compare and contrast her diaries from Angola to her public writings, such as her "handwritten circular letter" and its published form in *Mission Studies*. Arnott expressed concerns over her effectiveness in reaching her charges in Angola in her diary while her public writings were more positive. They were also able to compare her manuscripts to what was published and concluded:

> … In the places where her diction is revised, we see that the editing process honors her voice, content focus, and tone, opting for changes at the word (rather than the sentence or paragraph) level in ways that do not alter her overall message. One inference we might draw from comparing the two versions of this account is that Arnott's writing reached her magazine readers with only minor editorial interventions in between.
>
> Robbins and Ellis Pullen 2011, 272

Robbins and Ellis Pullen's work offers a glimpse into the editorial practices of missionary publications, revealing in at least one case the author's voice was largely preserved. Unlike male missionaries, female missionary authors had the added task of maintaining a balance between making the writing interesting and even lively while remaining a lady (Robbins and Ellis Pullen 2011, 77).

Just as the missionaries needed the support of the congregations "back home," missionaries provided an important element to Sunday schools and church meetings, adding a global and potentially imagination-capturing element as well as an education element. Missionary work and its publications were seen as an important and necessary supplement to the geographies being taught in American schools. In an essay published in *Mission Studies: Woman's Work in Foreign Lands*, a Mrs. R.S. Osgood argues for the importance of children learning about missionary work, for developing empathy for others as well as developing the general geographic knowledge:

> There is much in missionary study just as important for general information and as helpful to them for later work in school.
>   Geography is much more firmly fixed in their minds as they hear of Miss Clark in Africa, Miss Howe in Japan, Mrs. Coffing and Miss Bates in Turkey, Miss Burris in Mexico and all the rest of their own missionaries, in the different fields given to them to till.
>
> Osgood 1902, 219

Through their writings, missionaries were shaping the geographic imaginations of their audiences, reinforced by their promotion of cartographic culture.

"Make them if you cannot buy them ...": Mapping in support of mission work

> Use maps. Have at least the map of the world. *Make them* if you cannot buy them.
>
> Missions Studies, April 1902[9]

Maps were created and distributed in conjunction with missionary work for a variety of purposes. I am focusing here on widely available and circulated maps from missionary publications.[10] I am more interested in the broader circulation of maps through missionary publications and even more in the promotion of map consumption through mission studies. Through the use of and the advocacy of map use in mission studies, a practice of cartographic culture was being promulgated among American women and children.

In addition to their geographical writings, women missionaries visually represented geographic knowledge through a variety of means, employing photographs and maps to visually reinforce their mission work. Publication technology allowed for easy reproduction of photographs beginning in approximately 1890. Missionary publications, like all publications at the time, embraced photographs, and disseminated "visual images of exotic parts of the world" (Barringer 2004, 51–2). Between the photographs and the maps, they were constructing a visual geographic imagination.

The most common maps in women's mission publications were basic country reference maps (Figure 3.2), providing a sense of the country, major cities, and highlighting locations of missions. Women are credited with the preparation of a variety of these maps, such as the wall map of Southeast Asia by Annie Ryder Gracey's (Figure 3.3). In missionary publications, it was emphasized over and over again the importance of having at least a world map on hand when discussing missions (for example, Brain

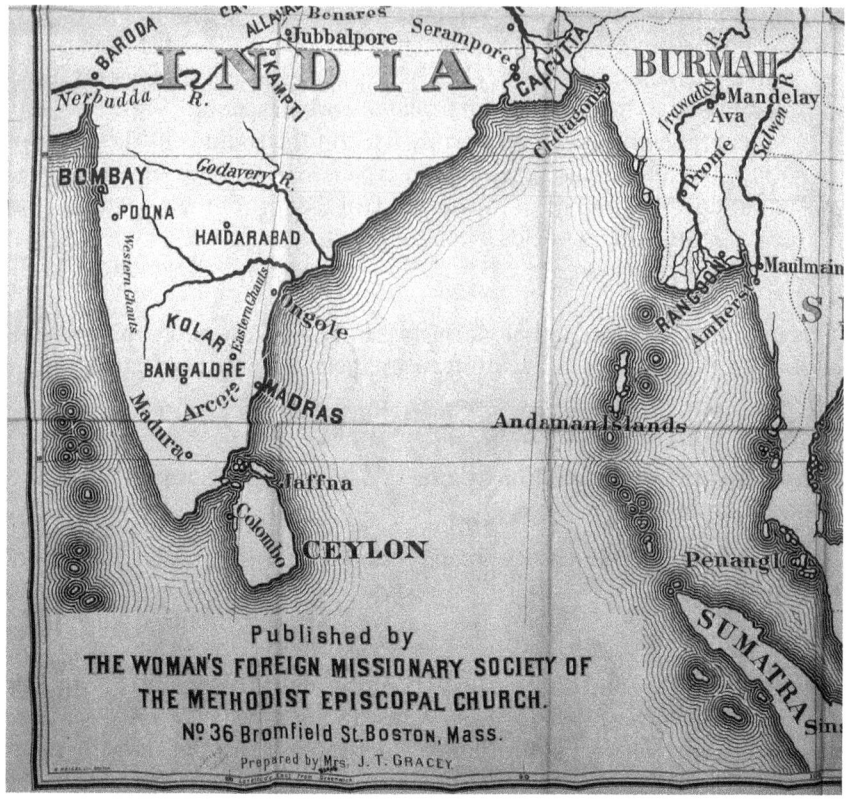

Figure 3.3 Detail of southwest quarter of [Map of Southern and central East Asia missionary stations] "prepared by Mrs. J.T. Gracey," circa 1880. A large wall map (132 cm × 164 cm), the map was manufactured to be used in Sunday school and women's study group meetings. By permission of The Newberry Library, Chicago, IL (folio D 806.569).[11]

1904, 128; Hixson 1906, 83). Articles, such as Mary Wainwright's, directed readers to look at the reference maps in the magazine issue and follow their journeys, linking the verbal descriptions to the spatial perspectives.

Less common were maps created to convey a sense of their own geographies, such as of the "Unitarian Settlements in North Carolina and Florida" or of the circuit traveled by a YWCA secretary in a southern U.S. county (Figures 3.4 and 3.5). These less sophisticated maps capture more of an immediate sense of the geographies of mission. In the cases of Figures 3.4 and 3.5, the rather crude maps were published in reports on women's home mission work. The map of Unitarian settlements is locating the communities where the Unitarians were working. The map appears to have been hand-drawn and not necessarily by a professional. In fact, the scale of the map is not effective for what they are trying to depict: small communities in eastern

Figure 3.4 Map from "Unitarian Settlements in North Carolina and Florida," 1915. The text states: "The question is often asked: Where is this work, and what is it? The communities are found in southeastern North Carolina, as will be seen by the map. If you are going to Shelter Neck, the train on the Atlantic Coast Line will leave you at a little town called Watha, about thirty miles north of Wilmington, and then there is a seven miles' drive through pine woods to the hamlet. For Shelter Neck is not a village, nor has it the distinction of a post office." By permission of the Wisconsin Historical Society, Madison, WI (Pamphlet 50-3587).

North Carolina and in the Florida panhandle. Its creator attempts to use numbers to identify the communities but on the published map the numbers are so small they are easy to overlook. The text does briefly make reference to the map. In the YWCA's "The Outlook for Town and Country" (1918), the brochure discusses the work of the YWCA to provide opportunities for girls in both urban and rural areas. While broadly discussing national efforts, the maps in Figure 3.5 depict "What the War Council is Doing in One Field," capturing efforts in the South Atlantic Field. The three maps use a variety of scales – states, counties around Atlanta, and then within a county. The maps are not discussed in the brief text. Captions provide some information about each map. However, not all information is provided. In the map to the left, no key is provided. In the upper right map, the lettering is almost too small to read. And in the lower right, the county and its state

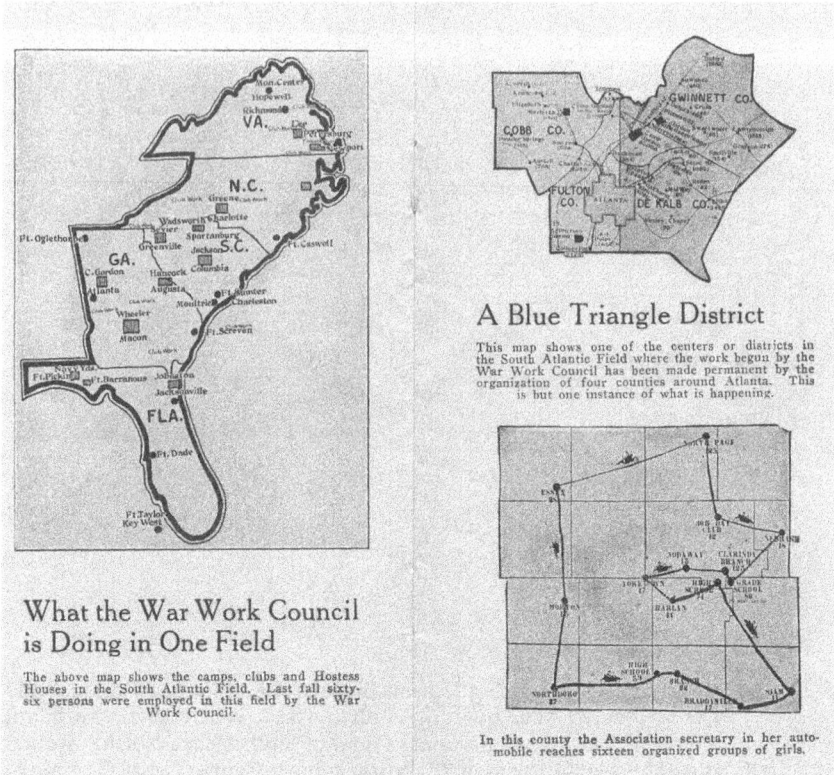

A Blue Triangle District

This map shows one of the centers or districts in the South Atlantic Field where the work begun by the War Work Council has been made permanent by the organization of four counties around Atlanta. This is but one instance of what is happening.

What the War Work Council is Doing in One Field

The above map shows the camps, clubs and Hostess Houses in the South Atlantic Field. Last fall sixty-six persons were employed in this field by the War Work Council.

In this county the Association secretary in her auto-mobile reaches sixteen organized groups of girls.

Figure 3.5 "The Outlook for Town and Country" published by the National Board Young Women's Christian Associations, 1918. The pamphlet talks in general about the efforts to provide "leadership and training and encouragement" to girls in both the town and the country. This illustration captures a two-page spread of maps of YWCA activities on a variety of scales, at the "South Atlantic Field," around Atlanta in the "Blue Triangle District," and then for an unidentified county. By permission of the Wisconsin State Historical Society, Madison, WI (Pamphlet 50-4431; Image ID 130590).

are never identified, items on the map (distances?) have been scratched out, and the lines have been crudely drawn. To me, this suggests perhaps these maps were created for another purpose – perhaps as part of a report from the South Atlantic Field? – and repurposed as illustrations in this pamphlet. Their readability is secondary to the visual power of maps.

In several instances, the maps accompanying mission reports from Africa lapse into propaganda. The association of Africa with "darkness" is visually captured in maps such as in Figure 3.6. In another example, directions for creating decorations for a Christmas "giving tree" specify the continent of

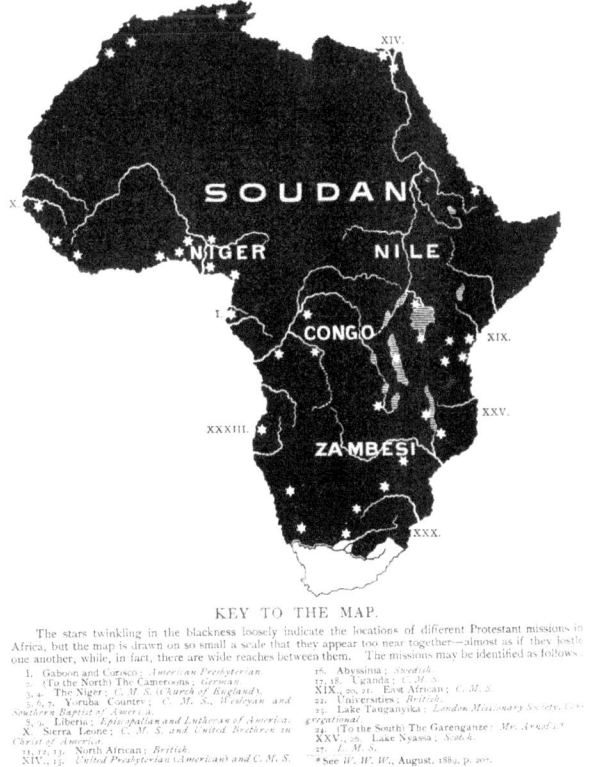

Figure 3.6 Africa as the "Dark Continent,": Protestant missions identified with "twinkling stars." Its caption reads "The stars twinkling in the blackness loosely indicate the locations of different Protestant mission in Africa, but the map is drawn on so small a scale that they appear too near together— almost as if they jostle one another, while, in fact, there are wide reaches between them." From *Woman's Work for Women*, June 1890, p. 148.

Africa was to be cut out of "black cardboard."[12] We are left to wonder how much they were aware of their propaganda-slant on Africa. Was the association of Africa with "darkness" so ingrained it was done without thought? Or was it a general practice that "mission stations conventionally appear as pinpoints of light in the heathen darkness" (Barringer 2004, 52)?[13] Yet the only examples I have found of such maps are of Africa.[14] Moral maps had been equating black or darkness with backwardness or lack of education and culture since at least the 1820s (Friendly and Palsky 2007, 240–2). Henry Morton Stanley, locater of David Livingstone and explorer in his own right, described Africa in two bestselling books as the "dark continent" – *Through the Dark Continent* (1878), and *In Darkest Africa* (1890). Or were such maps

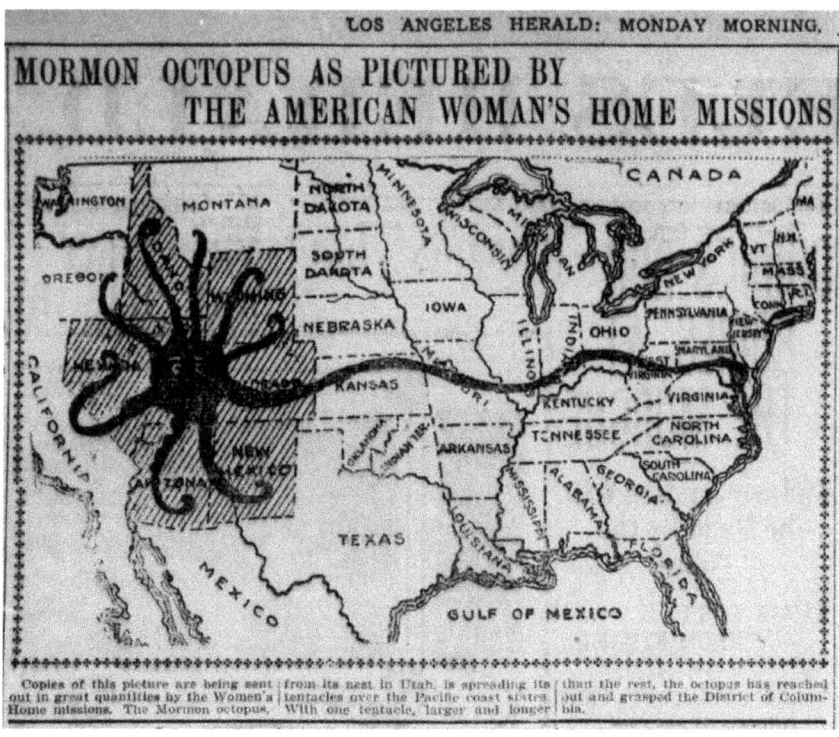

Figure 3.7 The "Mormon Octopus" first appeared in 1898. This version, with a tentacle reaching out to Washington, D.C., appeared in several publications. Shown here is from the *Los Angeles Herald* of 4 January 1904. It also appeared in an issue on missionary work in Utah of *Home Mission Monthly* dated October 1903.

of "darkest Africa" carefully constructed, employing widely understood cultural references with the intent of persuasion?

In a very consciously crafted propaganda map, the Mormon octopus map (Figure 3.7) was first published in a brochure from 1898 by the Woman's American Baptist Home Mission Society (*The Mormon Octopus* 1898). Over a basic map of the United States, a large black octopus has been positioned over Utah, the adjacent states under its tentacles tinted grey to indicate they are under its influence (suggestive of ink?). Octopi have been used on maps as threatening symbols of imperial powers since at least the 1870s and, in the United States, octopi were used to depict "'evil' combinations and political and corporate power" (Cannon 2013, 47; Tyner 2015, 1090). In several instance of editorial cartoons, Joseph Smith was depicted as an octopus (Cannon 2013, 10 and 58). A report in *Home Mission Monthly* describes how a large display version was created for a "Home Missionary exhibit" held at an Iowa church in 1903. Each Sunday school classroom was used as a booth representing a different mission location. The Mormon booth was judged "best for information" and featured:

> ... a large United States map, and pinned to it was an octopus with its arms extending to the various headquarters of the Mormon church in this country, and then beyond to all countries of Europe and Australia. The location of these headquarters was taken from a Mormon leaflet, 'Information for Tourists'... .
>
> Unique and Successful 1904, 272

The first Mormon octopus map had relatively short tentacles but over time the map was modified with the 1904 version adding a tentacle reaching out to Washington DC and a Canadian version adds tentacles reaching to Ottawa and Alberta (Woodworth 1909). The maps capture the high anxiety of non-Mormon Americans over the growth of the Mormon faith: "Like a huge octopus, the Mormon hierarchy is fastening its tentacles throughout the Rocky Mountain States, and is sapping the very life-blood of American freedom" (*The Mormon Octopus* 1898, 3).

In addition to the publication of maps, there was a great deal of discussion in mission publications of map use. Maps were presented as more than just reference materials employed during magazine reading. Writers promoted map consumption at all levels, from individual use to entire classes dedicated to them. In an essay on "The Missionary Periodical—Is It Read?" from 1909, the author makes the argument for the reading of missionary magazines, what can be gained through them, and how to read them. She writes: "Read with map before you, looking up all localities mentioned" (Raymond 1909, 120).

Directions for monthly mission studies meetings instructed the leaders to obtain maps and use them during their studies, suggesting a person be designated to point to the appropriate locations on the map during the lesson (Home Department ... 1916). In more than one instance, women were directed to make basic reference maps if they could not afford to purchase them (*Mission Studies* April 1902 and *Woman's Missionary Friend* March 1898). Some issues provided directions on how to make thematic maps to be used for the month's meeting, such as:

> The immigrant question can be graphically represented by the use of map that may be prepared by some member of the missionary society. The helps necessary for its development, which may be procured from headquarters, are as follows: An outline wall map of the United States (price 35 cents), little flag seals, one hundred in a box (10 cents), and picture medallions of Coming Americans (10 cents per sheet of 16 or 30 cents for 6 sheets).
>
> The ports by which the immigrants gain entrance to our land can be located by pasting one of the little flags at each point, and the states to

which they go can be shown by placing on each a little head cut from the picture sheet of the child representing the predominating nationality.... . This map when completed tells its own story.

Rue 1909, 80–1[15]

Some women's magazines provided monthly directions to the "Map Committee" on what maps would be needed the following month so they could be prepared.[16]

"Map talks" were another form of mission study, mini geography lessons prepared to accompany discussions of mission work in these countries, incorporating basic references and geographical facts.[17] Some are more dramatic and structured as a dialogue between a missionary and a friend or as a lecture by a native preacher. Initially directions were provided on map talks in the magazines (for example, Bowes 1914). A few were reprinted and distributed as pamphlets. At the back of many of the magazines were advertisements for publications and other "Necessities for Local Leaders": an October 1903 issue of *Home Mission Monthly* had a full page of publications and materials, including three map talk pamphlets: "Map Talk on Missions Among the Mormons," "Map Talk on Missions Among the Alaskans" and "Map Talk of Missions Among the Mountaineers," as well as a "Map Locating Mission Schools."[18] Eventually, a selection of map talks were compiled into the book *Making Missions Real* (1919). While Jay Stowell is credited with the publication, women are acknowledged as the authors of approximately half of the fifty-four "map talks" in the book.

The active engagement of women with maps ranges from active engagement with maps with educational aims in mind to performance where they are symbolicly referencing maps. In Martha Hixson's *Missions in the Sunday School*, Hixson provides a number of ways teachers can engage their students with maps, including using stars or flags to mark missions and running ribbons from missions to mission headquarters, or taking students on "imaginary journeys" using maps and photographs from books and magazines. Belle Brain in her book *Holding the Ropes: Missions in Sunday-School* writes of the importance of each Sunday-school having a world map and "The fields or stations to which the school has sent money should also be marked on the map, using gold stars or tiny flags for the purpose. This plan, used in Ralph Wells' school in New York City, greatly delighted the children, and had no small influence in increasing their gifts" (Brain 1904, 129). But active engagement edges towards performance when the students become the map. Hixson describes a map activity: "In one instance where conditions were favorable a map of the world was spread upon the floor and the children marked the stations with lighted candles while the teacher told the story of how the missionary money was used" (Hixson 1906, 87).

Performance cartography is "... a nonmaterial oral, visual, or kinesthetic social act, such as a gesture, ritual, chant, procession, dance, poem, story, or other means of expression or communication whose primary purpose is to define or explain spatial knowledge or practice" (Woodward and Lewis 1998, 4). The children marking the mission stations with lighted candles were essentially engaging in a symbolic performance ritually recognizing spatial locations. Through the practices of physically engaging with maps as part of a lesson (pointing at locations while someone speaks or marking locations with candles), or preparing maps for use in a lesson (crafting a thematic map), American women were actively engaging with the maps and with spatial data. In scholarship of education, it is widely acknowledged people learn in different ways. Some are visual learners who learn more effectively using graphics and/or symbolic ways of representing information, others are more kinesthetic learners who retain information when they are actively engaged with the material (Fleming and Mills 1992).

Maps have long been approached as concrete, static "truths" but increasingly, scholars are considering how maps are in-fact always in the process of becoming: "... they are brought into being and made to do work in the world" (Kitchin et al. 2012, 2). The practices of mapping, the active engagement with maps as manifested in women's missionary publications supports a practice of mapping as part of pedagogy. This process of using and making maps can be an important part of building community:

> Map-making is a potentially useful pedagogy because it can facilitate learners to understand, in spatial terms, how power claims can be asserted as truth and the effects that this has on everyday lives, as well as empowering them to spatially illustrate their own struggles and desires. Furthermore, the process of map-making can be at least as important as the produced maps, building collectively between participants.
>
> Firth 2014, 157

"Bring out your atlas ...": The collective geographical impact of missionary women

> Bring out your atlas, study up your geography, search your histories, delve into the encyclopedia... . These people afar off, these countries, circumstances, conditions, civilizations, are not myths, but realities: locate them, photograph them, realize them, and you will learn a new answer to the question, 'Who is my neighbor?'
>
> I.H. 1879, 38

Through their texts, maps, and calls to study geography and to map, missionary women were having an impact on the geographical imaginations of their readers. The publications produced by missionaries – such as the magazines and pamphlets – were designed for consumption in homes and in church spaces. Missionary publications not only published articles and lesson plans, they also highlighted articles in other missionary publications and regular magazines they deemed "Worth Reading."[19] At this point in U.S. history, Americans were curious about the world and there were relatively few sources of first-hand information outside of school textbooks and occasional magazine articles. *National Geographic* was just beginning.[20] Missionary publications provided geographic, ethnographic, even travel information about foreign lands to their audiences (Cooper 2002, 59; Jensz 2012, 248). Between the descriptions and photographs, they were essentially *National Geographic* before *National Geographic*. Mission presentations and exhibitions were another avenue through which geographic knowledge was produced and circulated, serving to entertain, educate but hopefully ultimately to motivate their audiences to support their work or perhaps even to become missionaries themselves (Linehan 2014, 437).

Georgina Endfield has written of the importance of the mission in the production of geographical knowledge, with the mission afforded a "scene" in "which existing geographical knowledge was applied and tested, and new knowledge, including that provided by local informants, was collected, assembled, exchanged and developed" (Endfield 2011, 203). Through their written accounts, missionaries combined accounts of geographies and cultures with conversion stories, creating "... an irresistible blend of the authoritative and the 'exotic' which may have been what rendered missionary writings so widely read" (Cooper 2002, 61). In the process, they were "enriching" Westerners' understanding of the world, both scientifically and on a broad public level (Livingstone 2005, 61).

The creation of geographic knowledge by women missionaries was largely constructing and circulating basic information on landscapes and cultures. Perhaps some did contribute to original geographic knowledge, like Matteo Ricci or David Livingstone. But for me, what is more significant is their production and circulation of information for other women, largely focused on "women's issues" – concerns over women's status and opportunities in these cultures and the lives of women and children in these countries. As a result, American women

> ... were familiar, if not conversant, with an entire vocabulary that implied the degradation of women: zenanas and harems; the seraglio and the bagnio; female infanticide and suttee; concubinage and polygamy; bride sale; foot-binding and ear and nose boring; consecrated prostitution and sacrifice; bastinado; child marriage and slavery.
>
> Brumberg 1982, 349

From this position of knowledge and experience, they were creating a reality for their audience. And this reality "... kept the global issues of the day in front of the eyes of American women, who felt special kinship with the Christian sisters they had helped education" (Robert 2002, 78).

Through the accounts constructed, audiences were taken on imaginative journeys. Sometimes it was literal, as the missionaries recounted step by step their journeys from the U.S. to their mission and their impressions along the way or it might be step by step through their mission town, such as with Mary Wainwright's account. These accounts helped their readers picture the location of the mission, but also linked the readers to the mission, "overcoming separations of geography and time to achieve mutual identification based on gendered, collaborative benevolence" (Robbins and Ellis Pullen 2011, 94). These missionary accounts were standing in for those who could not travel themselves, informing the geographical imaginations of those bound to home and church (McEwan 2000, 8).

Through the linkages nurtured – not only the articles but through letters and perhaps visits to US congregations – the missionaries and the U.S. congregations became part of singular, imagined communities. Missionaries were the "bridges between worlds—between West and East, North and South, the familiar and the exotic, the concrete and the spiritual" (Robert 2002, 60). Grounding these imagined communities were their shared beliefs but also "... a common literature, a common geography, a common time" as well as the missionaries themselves (Huber and Lutkehaus 1999, 21). Appeals for aid, assistance, and prayers were framed emotionally in the first person, creating an intimacy (Jensz 2012, 242). Missionaries "translated" these different cultures for their home supporters through their writings, describing the salient cultural characteristics and using their first-person narratives to make real to their readers the practices they were witnessing (Robbins and Ellis Pullen 2011, 25). It was making the "exotic" available for consumption yet appealing to their Christian ideals.

But this was a selected reality, constructed to appeal to their audience and encourage support for their undertakings. This basic geographic knowledge is filtered through the missionaries' lenses, not a "true" picture but selective, what they thought their audiences needed to know, so doubly selective from their perspective as a Westerner *and* filtered with the thought of educating as well as persuading their readers to support such work. While drawing attention to the plight of women in various parts of the world, they were "conveniently side-stepping the role and rights of women in the United States" and simultaneously holding the position of American women up as an exemplar, advocating not only for the missionaries to intercede but for the United States to intercede in these countries (Brumberg 1982, 367). While attempting to "uplift" the down-trodden of the world, their uplift was to their own standards of culture and society, importing essentially Western culture in the name of "civilizing." Ultimately, what is found is more a reflection of American Progressive Era ideals than anything else. Yet, they saw geographic knowledge as central to understanding the challenges of mission work in different counties.

Publications were the cornerstone of these efforts for they provided a form of "immutable mobile," that is, they are an object allowing the easy transmission of ideas and concepts while maintaining a certain degree of permanence, drawing from Actor-Network Theory (Latour 1986, 21). Similarly, in his analysis of two prominent economic textbooks, Trevor Barnes focuses on the ways in which texts perform: serving as an "inscription device," permitting ideas to diffuse without corruption; offering "optical consistency and semiotic homogeneity," allowing diverse phenomenon to be controlled and manipulated once it is reduced to the size of a page; becoming touchstones connecting networks; and representing powerful rhetoric, binding complex evidence and ideas into a cohesive whole, constructing compelling arguments (Barnes 2002, 493–5). Through the use of such immutable mobiles/inscription devices, mission publications brought together communities across the globe and, in the process, constructed a particular body of geographical knowledge.

In considering the construction of geographic knowledge, Charles Withers has thoughtfully noted:

> What we may be warranted in asserting as true, as rational, as legitimate natural or geographic knowledge is different from time to time and, even, from place to place.... Issues that have to do with *where* and *why* certain forms of knowledge took the discursive shape they did may, then, be inseparable from the socially constituted and locally credible matter of *how* they were produced.
>
> Withers 1999, 502[21]

Most histories of geography focus on the creation of geographic knowledge through the work of explorers, geographical societies, or scholars affiliated with universities. Withers and scholars such as Hayden Lorimer have argued we need to consider other contexts in which geographic knowledge was created and circulated and at a wider variety of scales, including local practice. Traditionally, only narrowly defined practices were recognized as being "real" geography but, in doing so, scholars failed to acknowledge the other arenas where geographic knowledge was generated and dispersed. Withers has suggested we need to widen our perspective as well as reconsider the construction of the history of geography, privileging certain practices and practitioners while discounting others (Withers 1999, 520). In Withers' work on the contributions of four educated men to the understanding of the geography of Scotland, he concludes "Making such geographical knowledge public depended, of course, on getting it published" (Withers 1999, 520–1). Missionary women were crafting, publishing, and distributing geographic knowledge for a particular audience and for a particular purpose. Only recently have scholars begun to consider their contributions to the corpus of geographic knowledge.

Thousands of American women took on the task of missionary work, both at home and abroad. Were they successful in their efforts? It's very hard to say. At the mission-level, converts were made at their missions, but seldom the number they hoped for, according to the accounts published. Hospitals and schools were built and at least for a period of time, efforts were made to educate women and girls and provide them with opportunities in medicine and education.

And their impact on geographic knowledge back at home? One minister wrote to one of the mission magazines, describing "I have here at the Saddle Mountain Church, a map of the world, and the United States, hanging on the wall, so that missions need not fail our thought."[22] In a published lecture, Helen Barrett Montgomery uses Abbie Gunn Baker of Washington DC as an example of how mission studies can foster a "world-mind" in Americans. Baker found herself seated next to a foreign attaché and chatted with him for some time about foreign affairs. He was impressed with her knowledge and asked how she obtained it. Baker attributed her knowledge to Mission Study classes. Surprised, the attaché remarked: "'This is the most interesting fact that I have learned while in America'" (Montgomery 1931, 17). Perhaps more telling are missionaries whose exposure to mission geography and maps led them to find their calling:

> Some of the greatest missionaries have received their life impulse from a study of maps. Eliza Agnew, when eight years of age, decided to become a missionary when her teacher in New York, which conducting a geography lesson, related to the class the story of a friend's life, and, placing his figure upon the Isle of France, told them not to forget that Harriet Newell was buried there; Alexander Mackay when a mere lad had his interest in missions intensified by tracing with his father upon the map of Africa the journeys of David Livingstone; and the mission of William Carey grew upon him as he studied the map of the world in giving lessons in geography in the little school at Moulton.
>
> Hixson 1906, 82–3[23]

Hixson includes the account in her chapter on "Map and Chart Work," providing these examples as evidence of the impact of including geography and map lessons as part of mission studies. While there is anecdotal evidence of the impact of mission studies, there is also its antithesis: frustrations voiced over not achieving fund-raising goals and concerns over apathy, with heaps of such magazines lying unread in churches (Barringer 2004, 52).

"We began in weakness, we stand in power"

> We began in weakness, we stand in power. In 1861 there was a single missionary in the field, Miss Marston, in Burma; in 1909, 1948 of them from the United States. In 1861 there was one organized woman's society in our country; in 1910 there were forty-four. Then the supporters numbered a few hundreds; to-day there are at least two millions. Then the amount contributed was $2000; last year four million dollars was raised... .
>
> Montgomery 1910, 243–4

... [S]ometimes the world they changed the most was the one that sent them.

Robert 2002, 81

During the Progressive Era, the most devoted supporters of missionary work were American women. Millions of American women, working through over forty mission societies, carried on local education, ran study groups, set up and ran children's groups, and "kept alive the interest in missions that was need to make them a vital part of church work instead of a remote and impersonal function" (Garrett 1982, 222–3). Through such work, American women were able to make a difference on a national as well as global level. Some scholars have argued women's missionary work can be seen as a form of feminism. While their goals superficially were to promote Christianity and "modern" Western social structures, their focus on education and the rights for women "struck at the heart of the cultures they entered" (Garrett 1982, 224). While working to "improve" the quality of life, it usually meant an end to cultural traditions upheld for centuries. Their focus on women's needs and addressing them through women's work can viewed as "the beginning of consciousness, the indisputable seeds of a woman's movement" (Garrett 1982, 228).

Yet, the surge in women missionaries was short-lived. Male-dominated boards took over most missionary societies in the 1920s and 1930s, and there was a significant decline in the publications, study groups, and other missionary support efforts (Reeves-Ellington 2011, 203). Missionaries continued and continue to spread their messages around the globe, but nowhere near the extent that the Progressive Era nurtured.

Cheryl McEwan, writing about Victorian women travelers, states:

> Women travelers may not have 'discovered,' mapped and explored 'new' territories, but they did add to the imagery constructed about the empire. Moreover, they influenced how this imagery was mapped onto the imaginations of the majority of people in Britain who did not have first-hand experience of areas of the world such as west Africa. The fact that they experienced places that many people in Britain could never experience, and that they conveyed impressions of these places to their readers, makes the work of these women 'geographic.'

McEwan 2000, 215–6

Similarly, American women missionaries added to the imagery constructed about the American spheres of influence, shaping the geographical imaginations of many Americans, especially women. I consider their work "geographic." It was, after all, "the chief means by which ordinary American church women gained information on non-Western religions, cultures, and women's issues around the world in the early twentieth century" and nurtured a global sisterhood (Robert 2002, 77). Women missionaries created this geographic knowledge for a specific audience and published it through a variety of means, sending it out to an audience across the United States.

I will now shift and take up our second strand of women's work: settlement houses.

Notes

1 The missionary explorer David Livingstone (1813–1873) is not to be confused with the contemporary historical geographer David N. Livingstone (1953–).

2 This figure of "over 500 alumnae engaged in missionary work" cited on the "Missionary Alumnae" page of the Mt. Holyoke website: https://www.mtholyoke.edu/archives/history/missionaries. Accessed 16 November 2015.

3 Jane Addams reportedly viewed the Moody Bible Institute, in the vicinity of Hull-House, with "disdain": Addams had "mixed feelings" about religion (Eddy 2010, 36).

4 The phrase "Bible women" was used both for American women missionaries and for converts to Christianity who assisted missionaries. Patricia Martin in her article "Ordained work – Unordained workers" suggests this phrase was used because "... to have simply called them ministers or deaconesses—even teachers or missionaries—would have challenged time-honored interpretations of the Bible and threatened the longstanding authority and power held by Baptist men. A new designation, one that did not coincide with the ordained positions reserved for men, avoided the issue and implied that the women would continue to operate within a separate, less-public sphere" (Martin 1988, 1–2).

5 When Jane Addams was in London in 1888, she visited the World Centennial of Foreign Missions, writing to her sister: "I have become quite learned on foreign missions and ashamed of my former ignorance." See http://www.janeaddamsproject.org/Jane Addams Draft_files/Page 3700.htm. (Accessed 8 March 2016).

6 I conducted a rudimentary content analysis on the articles, tabulating topics of discussion, ranging from aspects of regional geography to missionary concerns.

7 The results of my limited study align with the results of a published study: Chin 2003, see pp. 333–336.

8 From the January 1905 issue of *The Chinese Recorder and Missionary Journal* 36, 1.

9 Emphasis is from the original.

10 I am not addressing the maps created by missionaries as part of reports for their own ends, such as those discussed by Ruth Kark in "The Contribution of Nineteenth Century Protestant Missionary Societies to Historical Cartography" (Kark 1993).

11 The large size of the map made photography difficult and resulted in a detail standing in for the whole.

12 See *Lutheran Woman's Work* December 1917, p. 504.

13 Pellervo Kokkonen discusses a Finnish missionary map of Africa, dated 1878 and based on the work of Stanley, where "heathenism" is depicted in green and "Mohammedan" in grey with missionary stations as large white circles: "apparently representing the light of the Gospel shining widely through the surrounding darkness" (Kokkonen 1993, 162).

14 "The World in Boston" mission exposition, held in 1911, also featured "The Pageant of Darkness and Light" which features scenes of missionary work in North American reservations, Africa, India, and Hawaii (Ten Thousand ... 1911). "Darkness and Light" certainly fits with the "moral geography" of "heathen" lands as black or dark and Christianity as the light of redemption,

suggesting this was applied to more than just Africa. However, on maps, to date, the only examples I have located are of Africa or African countries.

15 Another excellent example provides directions about comparing/contrasting access to medical care in the United States versus China using maps of each. See *Mission Studies* November 1919, p. 308.

16 *Home Mission Monthly*, published by the Woman's Board of Home Missions of the Presbyterian Church, was one magazine that provided directions monthly to the "Map Committee," circa 1908–1909.

17 The first suggestion that a meeting could be constructed around a map dates from 1881 in *Heathen Woman's Friend* where they suggest drawing a simple map of the world on a chalkboard and "some lady" pointing out the locations of the mission stations and facts about each station and country (J.R.W. 1881).

18 The pamphlets were 3 cents each or $2.50 per hundred.

19 For an example, see *Women's Missionary Friend* 19, 5 (November 1897): 134.

20 Founded in 1888, *National Geographic* magazine began publishing photographs in 1896, and transitioned from a scholarly publication to a popular publication in around 1898 (Lutz and Collins 1993, 20–27).

21 Emphasis is Withers.

22 In the "From Our Far Lands" column, note from Harry H. Treat, Saddle Mountain Oklahoma in *Mission*, American Baptist international magazine, volume 1 page 426.

23 Hixson discusses these missionaries as part of her chapter on "Map and Chart Work" but she cites Belle Brain's *Holding the Ropes* (1904) as her source, pp. 63–64. In Brain, it is part of a discussion of how a wide variety of missionaries came to be called to this work.

Primary Sources

Abbey, L. 1905. Seeking good pearls in Nanking Field. *Woman's Work* 20, 2: 37–38.

A Great Field for Women. Chicago, IL: American Medical Missionary Society, 188–?

A survey of the field: ninety-nine years after Robert Morrison. 1906. *The Baptist Missionary Magazine* 86, 1: 9.

Bashford, J. 1905a. West China. *Woman's Missionary Friend* 37, 7: 238–239, 243.

———. 1905b. The awakening of China. *The Christian Advocate* 80, 8: 297–299.

Benn, R. 1912. *Ping-Kua: A Girl of Cathay*. Boston, MA: The Tudor Press for the Publication Office, Women's Foreign Missionary Society, Methodist Episcopal Church.

Bowes, L. 1914. Enlivening the lifeless society. *The Home Mission Monthly* 28, 6: 149–150.

Brain, B. 1904. *Holding the Ropes: Missionary Methods for Workers at Home*. New York & London: Gunk & Wagnalls Company.

Chandler, A. 1902. A vacation letter from Hakone. *Mission Studies: Women's Work in Foreign Lands* 20, 10: 301–2.

Dornblaser, I. 1911. China as seen by one of our own girls. *Lutheran's Woman's Work* 4, 9: 365–372.

Fensham, F. 1908a. Evangelism in the Far East, No. 1. *Mission Studies* 26, 1: 7–11.

———. 1908b. Evangelism in the Far East, No. 2. *Mission Studies* 26, 2: 44–48.

Fleming, E. 1920. A new memorial arch. *Woman's Work* 35, 8: 170–171.

Gilmore, D. 1905. The good old native preacher. *The Baptist Missionary Magazine* 85, 1: 31.

Gracey, L. 1893. *Gist: A Hand-Book of Missionary Information*. Cincinnati, OH: Cranston & Curts.

Hixson, M. 1906. *Missions in the Sunday School: A Manual of Methods.* New York, NY: Young People's Missionary Movement.

Home Department: Program for April Meeting, Subject: India. *Woman's Work* 31 (1916): 67.

I.H. 1879. Uniform readings: 'let there be light.' *Heathen Woman's Friend* 11 (Aug.): 37–39.

J.F.W. 1881. Home department – northwestern branch. *Heathen Woman's Friend* 12, 10: 234–235.

K.B.S. 1911. Children's department: "China special." *Lutheran Woman's Work* 4, 9: 373–374.

Missions and high life at Tengchow. 1905. *Woman's Work* 20, 2: 27–29.

Montgomery, H. 1910. *Western Women in Eastern Land: An Outline Study of Fifty Years of Woman's Work in Foreign Missions.* New York: The Macmillan Company.

———. 1931. *The Preaching Value of Missions.* Philadelphia, PA: The Judson Press.

The Mormon Octopus. Boston, MA: Woman's American Baptist Home Mission Society, 1898.

Mormon Octopus as pictured by the American Woman's Home Missions. *Los Angeles Herald,* 4 January 1904.

News of the year from girls' schools. 1905. *Woman's Work* 20, 2: 33–35.

North, L. 1897. Modern 'schools of prophets.' *Woman's Missionary Friend* 29, 5 (November): 126–130.

Ogburn, K. 1906. A school celebration at Nanchang. *Woman's Missionary Friend* 38, 5: 164–165.

'Open door' in the East. 1906. *The Baptist Missionary Magazine* 86, 1: 3.

Openshaw, H. 1905. A loud call for advance: West China conference voices a great need. *The Baptist Missionary Magazine* 85, 8: 305–306.

Osgood, R. 1902. Defrauding the children. *Mission Studies: Woman's Work in Foreign Lands.* 20, 7: 219–221.

Our four fields: the year. 1906. *The Baptist Missionary Magazine* 86, 1: 13–14.

The Outlook for Town and Country. 1918. New York NY: National Board, Y.M.C.A.

Parmalee, F. 1902. Christian work in Matsuyama, Shikoku. *Mission Studies: Women's Work in Foreign Lands* 20, 10: 310–312.

Raymond, M. 1909. The missionary periodical—is it read? *Home Mission Monthly* 23, 5: 118–121.

Rue, S. 1909. January aids. *Home Mission Monthly* 23, 3: 80–81.

Rugh, G. 1918. A vacation in China. *Lutheran Woman's Work* 11, 4: 119–121.

Snodgrass, M. 1905. Missions and high life in Tengchow. *Woman's Work* 20, 2 (February): 27–29.

Stowell, J. 1919. *Making Missions Real: Demonstrations and Map Talks for Teen Age Groups.* New York: The Abingdon Press.

Ten Thousand People to Portray Missionary Life. *The New York Times* 22 January 1911.

The Annual Report. 1905. *China's Millions* 17, 7: 73–77.

Unique and Successful. 1904. *Home Mission Monthly* 18, 11: 272–273.

Wainwright, M. E. 1902. Okayama. *Mission Studies: Women's Work in Foreign Lands* 20, 10: 294–296.

Welpton, C. 1902. From the mountain tops. *Mission Studies: Women's Work in Foreign Lands* 20, 10: 304–305.

The Woman's College in North China. 1920. *Woman's Work* 35, 8: 178–180.
Woodworth, J. 1909. *Strangers within Our Gates, or Coming Canadians.* Toronto: The Missionary Society of the Methodist Church, Canada.
Wright, M. 1897. The land of Li Hung Chang. *Mission Studies* 2 (March): 79–82.

Bibliography

Barnes, T. 2002. Performing economic geography: two men, two books, and a cast of thousands. *Environment and Planning A* 34: 487–512.
Barnett, C. 1998. Impure and worldly geography: the Africanist discourse of the Royal Geographical Society, 1831–1873. *Transactions of the Institute of British Geographers* NS 23: 239–251.
Barringer, T. 2004. What Mrs. Jellyby might have read. Missionary periodicals: a neglected source. *Victorian Periodicals Review* 37, 4: 46–74.
Bederman, G. 1989. 'The women have had charge of the church work long enough": The Men and Religion Forward movement of 1911–1912 and the masculinization of middle-class Protestantism. *American Quarterly* 41, 3: 432–465.
Braun, L. 2012. Missionary cartography in Colonial Africa: cases from South Africa. In *History of Cartography, Lecture Notes in Geoinformation and Cartography 6*, eds. E. Liebenberg and I.J. Demhardt, pp. 249–272. Berlin: Springer-Verlag.
Bridges, R. 2008. The Christian vision and secular imperialism: missionaries, geography and the approach to East Africa, c.1844–1890. In *Converting Colonialism: Visions and Realities in Mission History, 1706–1914*, ed. D. Robert, pp. 43–60. Grand Rapids: William B. Eerdmans Publications Company.
Brumberg, J. 1982. Zenanas and girlless villages: the ethnology of American Evangelical women, 1870–1910. *The Journal of American History* 69, 2: 347–371.
Cannon, K. 2013. 'And now it is the Mormons': the magazine crusade against the Mormon Church, 1910–1911. *Dialogue: a Journal of Mormon Thought* 46, 1: 1–62.
Chin, C. 2003. Beneficent imperialists: American women missionaries in China at the turn of the twentieth century. *Diplomatic History* 27, 3: 327–352.
Cooper, J. 2002. The invasion of personal religious experiences: London Missionary Society missionaries, imperialism, and the written word in early 19th-century southern Africa. *Kleio* 34, 1: 49–71.
Cox, J. 2008. What I have learned about missions from writing The British Missionary Enterprise since 1700. *International Bulletin of Missionary Research* 32, 2: 86–87.
Cramer, J. 2003. White womanhood and religion: colonial discourse in the U.S. women's missionary press, 1869–1904. *The Howard Journal of Communications* 14: 209–224.
———. 2004. Cross purposes: publishing practices and social priorities of nineteenth-century U.S. missionary women. *Journalism History* 30, 3: 123–130.
Deegan, M. 1988. *Jane Addams and the men of the Chicago School.* New Brunswick NJ: Transaction Books.
Dries, A. 2006. American Catholic women missionaries, 1870–2000. In *Encyclopedia of Women and Religion in North America*, eds. R. Keller and R. Ruether, pp. 843–850. Bloomington, IN: Indiana University Press.
Driver, F. 1999. *Geography Militant: Cultures of Exploration and Empire.* Oxford: Wiley-Blackwell.
Eddy, B. 2010. Struggle or mutual aid: Jane Addams, Petr Kropotkin, and the progressive encounter with social darwinism. *The Pluralist* 5, 1: 21–43.

Edwards, W. and C. Gifford. 2003. Introduction: restoring women and reclaiming gender in social gospel studies. In *Gender and the Social Gospel*, eds. W. Deichmann Edwards and C. De Swarte Gifford, pp. 1–17. Urbana and Chicago: University of Illinois Press.

Endfield, G. 2011. The mission. In *The SAGE Handbook of Geographical Knowledge*, eds. J. Agnew and D. Livingstone, pp. 202–216. London: SAGE Publications.

Firth, R. 2014. Critical cartography as anarchist pedagogy? Ideas for praxis inspired by the 56a infoshop map archive. *Interface: A Journal for and About Social Movements* 6, 1: 156–184.

Fleming, N. and C. Mills. 1992. Not another inventory, rather a catalyst for reflection. *To Improve the Academy*. Paper 246. http://digitalcommons.unl.edu/podimproveacad/246. Accessed 1/18/16.

Flemming, L. 1989. Introduction: studying woman missionaries in Asia. In *Women's Work for Women: Missionaries and Social Change in Asia*, ed. L. Flemming, pp. 1–10. Boulder, CO: Westview Press.

Flynt, W. and G. Berkley. 1997. *Taking Christianity to China: Alabama Missionaries in the Middle Kingdom, 1850–1950*. Tuscaloosa and London: The University of Alabama Press.

Fraley, J. 2011. Missionaries to the wilderness: a history of land, identity, and moral geography in Appalachia. *Journal of Appalachian Studies* 17, 1 & 2: 28–41.

Friendly, M., and G. Palsky. 2007. 'Visualizing Nature and Society.' In *Maps: Finding Our Place in the World*, eds. J. Akerman and R. Karrow, pp. 207–253. Chicago, IL: University of Chicago Press.

Garrett, S. 1982. Sisters all. Feminism and the American women's missionary movement. In *Missionary Ideologies in the Imperialist Era*, eds. T. Christensen and W. R. Hutchinson, pp. 221–230. Denmark: Aros.

Gifford, C. 2006. Nineteenth- and twentieth-century Protestant social reform movements in the United States. In *Encyclopedia of Women and Religions in North America*, eds. R. Keller and R. Ruether, pp. 1021–1038. Bloomington, IN: Indiana University Press.

Hasinoff, E. 2010. The missionary exhibit: a frustration and a promise for Franz Boas and the American Museum of Natural History. *Museum History Journal* 3, 1: 81–102.

Hill, P. 1985. *The World Their Household: The American Woman's Foreign Mission Movement and Cultural Transformation, 1870–1920*. Ann Arbor, MI: The University of Michigan Press.

Huber, M. and N. Lutkehaus. 1999. Introduction: gendered missions at home and abroad. In *Gendered Missions: Women and Men in Missionary Discourse and Practice*, eds. M. Huber and N. Lutkehaus, pp. 1–38. Ann Arbor, MI: University of Michigan Press.

Jensz, F. 2012. Origins of missionary periodicals: form and function of three Moravian publications. *Journal of Religious History* 36, 2: 234–255.

Kark, R. 1993. The contribution of nineteenth century Protestant missionary societies to historical cartography. *Imago Mundi* 45: 112–119.

Kitchin, R., J. Gleeson and M. Dodge. 2012. Unfolding mapping practices: a new epistemology for cartography. *Transactions of the Institute of British Geographers* NS 38, 3: 480–496.

Klein, T. 2013. Media events and missionary periodicals: the case of the Boxer War, 1900–1901. *Church History* 82, 2: 399–404.

Kokkonen, P. 1993. Religious and colonial realities: cartography of the Finnish Mission in Ovamboland, Namibia. *History in Africa* 20: 155–171.

Langmore, D. 1982. A neglected force: white women missionaries in Papua 1874–1914. *The Journal of Pacific History* 17, 3: 138–157.

Latour, B. 1986. Visualization and cognition: drawing things together. *Knowledge and Society: Studies in the Sociology of Culture Past and Present* 6: 1–40.

Lindley, S. 2006. The social gospel. In *Encyclopedia of Women and Religion in North America*, eds. R. Keller and R. Ruether, pp. 1069–1076. Bloomington, IN: Indiana University Press.

Linehan, D. 2014. Irish Empire: assembling the geographical imagination of Irish missionaries in Africa. *Cultural Geographies* 21, 3: 429–447.

Livingstone, D. 2005. Scientific inquiry and the missionary enterprise. In *Participating in the Knowledge Society: Researchers Beyond the University Walls*, ed. R. Finnegan, pp. 50–64. Basingstoke: Palgrave Macmillan.

Lutz, C. and J. Collins. 1993. *Reading National Geographic*. Chicago, IL: University of Chicago Press.

Martin, P. 1988. Ordained work—Unordained workers: Texas "Bible women," 1880–1920. *Texas Baptist History: Journal of the Texas Baptist Historical Society* 8: 1–9.

McEwan, C. 2000. *Gender, Geography and Empire: Victorian Women Travellers in West Africa*. Aldershot UK: Ashgate.

Ogawa, M. 2004. 'Hull House' in downtown Tokyo: the transplantation of a settlement house from the United States into Japan and the North American missionary women, 1919–1945. *Journal of World History* 15, 3: 359–387.

Reeves-Ellington, B. 2011. Women, Protestant missions, and American cultural expansion, 1800–1938: a historiographical sketch. *Social Sciences and Missions* 24: 190–206.

Robbins, S. and A. Ellis Pullen. 2011. *Nellie Arnott's Writings on Angola, 1905–1913: Missionary Narratives Linking Africa and America*. Anderson, SC: Parlor Press.

Robert, D. 2002. The influence of American missionary women on the world back home. *Religion and American Culture: A Journal of Interpretation* 12, 1: 59–89.

———. 2006. Protestant women missionaries: foreign and home. In *Encyclopedia of Women and Religion in North America*, eds. R. Keller and R. Ruether, pp. 834–843. Bloomington, IN: Indiana University Press.

Rynbrandt, L. 1998. Caroline Bartlett Crane and the history of sociology: salvation, sanitation, and the social gospel. *The American Sociologist* 29, 1: 71–82.

Singh, M. 2005. Women, mission, and medicine: Clara Swain, Anna Kugler, and early medical endeavors in Colonial India. *International Bulletin of Missionary Research* 29, 3: 128–133.

Taylor, S. 1979. The sisterhood of Salvation and the sunrise kingdom: congregational women missionaries in Meiji Japan. *Pacific Historical Review* 48, 1: 27–45.

Tyner, J. 2015. Persuasive cartography. In *History of Cartography* Volume 6, ed. M. Monmonier, pp. 1087–1095. Chicago, IL: University of Chicago Press.

Williams, J. and V. Maclean. 2012. In search of the kingdom: the social gospel, settlement sociology, and the science of reform in America's Progressive Era. *Journal of the History of the Behavioral Sciences* 48, 4: 339–362.

Withers, C. 1999. Reporting, mapping, trusting: making geographical knowledge in the late seventeenth century. *Isis* 90: 497–521.

Woodward, D. and G. Lewis. 1998. Introduction. In *History of Cartography Volume 2, Book 3*, eds. D. Woodward and G. Lewis, pp. 1–10. Chicago, IL: University of Chicago Press.

4 "Woman's work" Part II – Hull-House and social settlement work

> For obviously such work is woman's work; it has for its very essence the power of home-making, which has always been supposed to be a feminine prerogative.
>
> Scudder 1890, 10

A different form of woman's work can be found in the social settlement work of the Progressive Era. Vida Scudder, writing in 1890 about college settlements, was arguing for higher education for women, as settlement work as a logical application of women's education, and "settlement work was naturally the purview of women" (Williams and MacLean 2015, 52–3). Beginning in England, and diffusing to the United States, settlement houses began as a means of addressing social concerns tied to urbanization, industrialization, immigration, and poverty. They were termed "settlement" houses because their residents "settled" in neighborhoods to bring about significant social reforms (Barbuto 1999, vii). Toynbee Hall in London was the first to attempt communal living combined with social activism, opening in 1884 in Whitechapel. Jane Addams and Ellen Gates Starr were inspired by Toynbee Hall and looking for ways to use their educations and have meaningful lives without becoming missionaries. Their Hull-House, opened in 1889, became the largest and most well-known of the social settlements in the United States, but there were social settlement houses across the country, in all major cities and even in mid-size cities.[1]

The majority of workers were women, largely white, Protestant, and middle class with many having at least some college education (Stebner 2006, 159). For some, settlement work was an alternative to religious missionary work, while other settlement workers followed the social gospel and found meaningful work in the settlement houses. Settlement houses took a variety of forms:

University settlements – founded and operated by university men.[2]
College settlements – founded and operated by college women.

Church settlements – founded and funded by particular congregations or
denominations.
Social settlements – non-denominational, but the term was also used as a
catchall for all of the above.

Individuals who lived and worked at settlement houses were termed "residents." It is difficult to get exact figures on the number of settlement
houses, as some early counts excluded certain religious groups and certain
socio-economic groups. But conservative figures suggest a growth of settlement houses from three in 1890 to over 400 by 1910, with 1,077 women residents and 322 men (Lengermann and Niebrugge-Brantley 2002, 6; Stebner
2006, 1061). In 1911, there were settlement houses not only across the United
States but also in Canada, Belgium, France, Germany, Holland, Italy, Swit-
zerland, and Japan (Ogawa 2004; Woods and Kennedy 1911, 305–309).

The concept of the settlement houses was based on settlement workers
helping others while helping themselves. By living in poorer neighborhoods,
working with the people, it was believed the settlement workers would "see"
or determine what the community needed, while having an opportunity to
contribute in a meaningful way to society (Stebner 2006, 1061). Some scholars argue settlement workers came the closest to truly achieving the goals of
the social gospel – "… by living among the working people and becoming
good neighbors, thereby putting into practice the social gospel admonition
of loving one's brother as oneself" (Williams and MacLean 2012, 344–5).
Most settlement houses had head residents and then anywhere from a couple to as many thirty residents, who all worked at the houses while paying
room and board. Settlement workers included many women who had attended college or university.

The "work" carried out by the settlement houses varied according to the
house, location, and the community needs, but activities common at many
of the houses included a wide variety of clubs for men, women, and children,
lectures and educational opportunities, and gymnasiums and occasions for
physical activity. They might also offer employment and relief bureaus, medical clinics and health programs, day care or kindergartens, and public playgrounds. Many settlement houses, in order to better serve their communities,
carried out community studies, conducting some of the first research in poverty, unemployment, living conditions, and addictions (Stebner 2006, 1062–3).

I will begin by focusing on Hull-House; not only is it the "queen" of the
settlement houses, but they also promoted the practices of social studies
and mapping. I will then consider the practice and promotion of geography
and mapping through social settlements' "social surveys." Similar practices
were also used by organizations with ties to social settlements. Finally, I will
explicate the ties between the nascent field of geography and Addams and
Hull-House. Through social settlements efforts to know and support their
communities they were engaging in practices of what we would today call
"social geography."

Hull-House: Settling and investigating

> To provide a center for the higher civic and social life; to institute and maintain education and philanthropic enterprises; and to investigate and improve the conditions in the industrial districts of Chicago.
>
> Hull-House Charter (quoted in Woods and Kennedy 1911, 53)

In the Hull-House charter, Addams, Starr, and residents establish their goals of creating a community center, providing a base for education and philanthropic activities, and studying and improving the conditions of their neighborhood.[3] As one of the most famous social settlements in the world, housing one of the most important woman activists in American history, Jane Addams, Hull-House's charter does indeed reflect their actions. Their goal was to become the center of a web of activity on Chicago's South Side, which is indeed what they achieved: bringing in community neighbors for clubs, classes, and activities; providing space for organizations to become established, such as the Illinois chapter of the National Women's Trade Union League, and offering educational opportunities; and conducting investigations into a wide variety of problems and working on addressing these problems. Hull-House became a key hub in Chicago, linking many individuals who lived and worked in the city, but also reaching much further ... across the United States and beyond. Major global activists interested in social issues visited Hull-House, including William Stead, Patrick Geddes, and Pyotr Kropotkin. Addams was acquainted with not only Stead and Kropotkin but also Leo Tolstoy and Mahatma Gandhi (Maude 1902).[4] In the course of their work, Hull-House residents practiced geography, demonstrating how place-based investigations and spatial analysis can illuminate social issues, disseminating these concepts throughout their networks, creating a culture of spatial analysis amongst Progressive Era activists, particularly women activists.

Founded by Jane Addams and Ellen Gates Starr in 1889, Hull-House was inspired by the social improvement efforts of Toynbee Hall in London. Addams and Starr had met at Rockford Female Seminary and were looking for opportunities to put their education to use. Addams visited Toynbee Hall in June 1888 and saw a possible model for their work. Addams and Starr presented their plans to leaders of the Chicago Woman's Club and received their support for their venture (Schultz and Hast 2001, xxxiii). Addams and Starr believed "by living with working-class and poor immigrants and sharing daily experiences they could find solutions to societal problems growing out of social conditions in slums" (Schultz 2007, 2).

Hull-House, located at the intersection of Polk and Halstead Streets in Chicago, was surrounded by tenements filled with immigrants and sweatshops, mixed with factories, warehouses and small neighborhood businesses (Horowitz 1983, 42). Edith Abbott described the Hull-House neighborhood:

> The foreign colonies were well established, and there were Italians in front of us and to the right of us; and to the left a large Greek colony.

There was a Bulgarian colony a few blocks west of Halsted Street and along to the north that had almost no women; but large numbers of fine Bulgarian men seemed to have emigrated – and they were pitiful when they were unemployed. Then you came to the old Ghetto as you followed Hull-House a few blocks to the south where the Maxwell Street Market with its competing pushcarts heaped with shoes, stockings, potatoes, onions, old clothes, new clothes, dishes, pots and pans, and food for the Sunday trade was as picturesque as it was insanitary.

Abbott 1950, 378

Eventually, Hull-House would comprise not only the former Hull family home but twelve additional buildings. A 1907 *Hull-House Yearbook* states: "Six thousand people come to Hull-House each week during the winter months, either as members of organizations or as parts of an audience" (Hull-House 1907).

While neighborhood needs were unquestionably great, part of Addams and Starr's vision of Hull-House was to find meaningful work for themselves and others like them: well-educated women who had few job prospects given society's restrictive roles for women at the time (Sklar 1985, 663–4). Individuals would come and stay for a period at Hull-House and then move on to other projects while maintaining their ties to Hull-House. Others stayed and never left. Agnes Holbrook wrote of Addams and Starr: "They came for good. They threw in their lot with the nineteenth ward, to reside, to 'settle,' not to experiment or test" (Holbrook 1894, 177).

Chicago was going through tremendous changes in urban life as it rapidly expanded, from a population of about 500,000 in 1880 to over 2.5 million people in 1920 (Diner 1980, 52). Immigrants were pouring into the city and finding work in the Stockyards, mills, railroads, and sweatshops. Whole families worked: fathers, mothers, and children. There were few if any labor laws regulating child labor, length of the work day, and working conditions. Living conditions were also difficult: poor sanitation, poor housing conditions, poor diets. Beginning in the 1870s, Chicago women activists worked to advance two agendas: "the advancement of white, middle-class women into the economic mainstream, and a reform agenda that had as a major goal the protection and assistance of working women and children" (Schultz and Hast 2001, xxv). With these goals in mind, Chicago women worked on many fronts to improve their city: improving schools and educational opportunities; advocating for pure food and drug legislation; improving city sanitation (sewage, street cleaning, garbage removal); founding hospitals and clinics; addressing juvenile delinquency and establishing juvenile courts; campaigning against venereal diseases and prostitution (Schultz and Hast 2001, xxxiii). Hull-House became a hub for many of these efforts, providing meeting space, connecting like-minded individuals, and coordinating efforts.

As Addams and Gates began their efforts in the Chicago South Side community, community investigations were integral to their work. In this, they were inspired by early social work out of London, both in living in the communities they were trying to elevate and in the practice of investigations.[5] In Agnes Holbrook's introduction "Map Notes and Comments," she acknowledges the inspiration of Charles Booth's maps of London on this work, with Booth's work conducted out of Toynbee Hall.

Charles Booth's work in London was driven by his desire to understand: who were the poor and why were they poor (Bales 1991, 68)? Booth was a wealthy Liverpool ship owner, merchant and manufacturer (Bales 1991, 66; Bulmer et al. 1991, 19). London was deep in a public debate over poverty but there was little factual information available. Booth sought to remedy this by gathering data on the citizens of London and levels of poverty and proposing answers to such social problems. Booth's maps of London poverty began with data collected by the School Board Visitors (essentially truant officers) who generalized the average income for each city block in London, depicting the location and level of poverty (Sklar 1991, 123). Booth and his associates supplemented the School data with census data, and additionally visited every street (Kimball 2006, 356–7). Using this data, they assigned each block a class, producing a map where the blocks were colored according to the conditions of households. The first map of East London was published in 1889. According to Topalov, "This was probably the first thematic map to cover an entire city together with the spatial distribution of the 'social conditions' of its population" (Topalov 1993, 408). Four maps covering all of London were published in 1891 and twelve maps covering all of London in 1902 (Kimball 2006, 358). Booth used colors of pink and red for "ordinary earning" and "highly-paid labor" and black and dark blue for "lowest class" and "casual earning" (poverty-level); in doing so, he presented poverty as not as extensive as thought, rendering it in manageable scale: "What had seemed visually chaotic has now become clear, organized and transparent" (Kimball 2006, 360 and 362). With such easily graspable information, action could be taken, solutions formulated: "Contemporary viewers commonly heralded the maps as a clear, 'bird's-eye-view' of the problem" (Kimball 2006, 359). Booth would eventually propose a "universal old age pension" and the "expansion" of the city (migration of "better-off" workers to the periphery) (Topalov 1993, 403). But this visual data was designed for display and Booth made certain it was widely seen: displayed at Toynbee Hall and Oxford House in 1888; the Royal Statistical Society in 1888; and the Paris international exhibition in 1900 (Kimball 2006, 359; Topalov 1993, 396 and 418).[6]

Like Booth's work, the research Hull-House conducted was done in the name of improving the quality of life in this community as well as providing intellectual engagement and meaningful work for the residents. The investigations Hull-House is best known for are the maps and studies published in *Hull-House Maps and Papers* (1895) but they represent but one of many investigations carried out by Hull-House residents. Residents used scientific methods to conduct fieldwork, tabulated statistical data, analyzed the data and created graphs, charts, and maps to display the data, then formulated results and presented possible solutions to the social problems they were observing (Grant et al. 2002, 70; MacLean and Williams 2012, 239–40). They produced a wide range of studies, from milk contamination to ethnic enclaves to workers' conditions in the sweatshops. These studies may be considered public geography with their spatial focus on the conditions in neighborhoods and many using maps to visually display their data and support their arguments. A 1911 overview of Hull-House identified fourteen studies conducted by Hull-House, including two for the U.S. Department of Labor and two for the U.S. Department of Agriculture (Woods and Kennedy 1911, 54). Among Jane Addams' papers are twenty-three "investigations and research," dating from 1892 to 1933 (Bryan et al. 1996, 36). I have located thirty-seven investigations among various Hull-House papers and do not consider my list definitive by any means (Table 4.1).[7]

While today *Hull-House Maps and Papers* is widely known and singled out for their ground-breaking work, it is but the tip of an iceberg of research conducted out of Hull-House. In fact, agencies came to Addams and residents for such investigations, knowing of Hull-House residents' knowledge of the neighborhood and that the neighborhood's familiarity with the residents facilitated such work. As Lengermann and Biebrugge-Brantley so aptly put it: "The settlement residents' most powerful weapon was their cultural capital" (Lengermann and Niebrugge-Brantley 2002, 11). When the results of the studies were published, many identified the authors/researchers as "of Hull-House" (such as Atwater and Bryant 1898; Britton 1906). Reflecting on a dietary study Hull-House carried out for the U.S. Department of Labor, A.C. True, Director of the U.S. Department of Agriculture's Office of Experiment Stations, wrote: "The familiarity which the Hull-House residents have with the conditions existing in that region of Chicago in which their settlement is located, and the confidence which they have inspired in the people for whose good they have labored, were important factors in the successful management of the investigations" (Atwater and Bryant 1898, 3). Hull-House contributions were significant enough to the study that it states on the title page, "Conducted with the cooperation of Jane Addams and Caroline L. Hunt, of Hull-House," listed above Atwater and Bryant's names and in larger, all-capitals type. In considering Hull-House investigations, I will begin with *Hull-House Maps and Papers* (hereafter *HHMP*) before considering the breadth of Hull-House investigations.

In 1893, resident Florence Kelley had been contracted by the Department of Labor to study a "slum" of Chicago: she chose the neighborhood to the southeast of Hull-House, a "port of entry neighborhood" and a "quintessential slum" (Schultz 2007, 6–7). As the "schedule men" turned in their completed schedules or forms to Florence Kelley, the information would be copied by a resident before being sent to the Commission of Labor in Washington DC (Schultz 2007, 18). Kelley and Agnes Holbrook organized the data and made the maps, color-coding the data and transferred onto maps provided by Chicago surveyor Samuel Greeley (Residents of

Table 4.1 Hull-House Investigations

Year	Investigation	Residents involved	Published as	Notes
n.d.	Tramps	Unidentified		Memo to J. Addams in *JAP Microfilm*
1892	"Investigations into the Sweating System for the State Bureau of Labor Statistics"	Florence Kelley	*Hull-House Maps and Papers* (1895); First, Second, Third, and Fourth Annual Reports of the Factory Inspectors of Illinois (1894, 1895, 1896, 1897)	Listed in: *Handbook of Settlements* (1911)
1892	School conditions	Florence Kelley, Corinne Brown	Brochure published by the Illinois Woman's Alliance (1892)	In *JAP Microfilm*
1893	"The Slums of Great Cities for Department of Labor"	Florence Kelley	C. Wright, *The Slums of...* (1894); *Hull-House Maps and Papers* (1895)	Listed in: *Handbook of Settlements* (1911); Howe 1896, 120.
1893?	Conditions of the Cloakmaking Trade in Chicago and New York	Isabel Eaton	*Hull-House Maps and Papers* (1895)	
1893?	The Chicago Ghetto/Jewish Population of Chicago	Charles Zeublin	*Hull-House Maps and Papers* (1895)	
1893?	Bohemian People in Chicago	Josefa Humpal-Zeman	*Hull-House Maps and Papers* (1895)	
1893?	Italian Colony in Chicago	Alessando Mastro-Valerio	*Hull-House Maps and Papers* (1895)	
1893?	Cook County Charities	Julia Lathrop	*Hull-House Maps and Papers* (1895)	
1893?	Dietary Investigation for Department of Agriculture	Ellen H. Richards, Amelia Shapleigh	A. Shapleigh, "A Study of Dietaries" (1894); R. Milner, ed., *Dietary Studies in Boston...* (1903)	Listed in: *Handbook of Settlements* (1911). In *JAP Microfilm*
1896	"Investigation of the Saloons of the Nineteenth Ward for the Committee of Fifty"	Ernest Carroll Moore	E. Moore, "The Social Value of the Saloon" (1897b); Included in J. Koren, *Economic Aspects of the Liquor Problem* (1899)	Listed in: F. Kelley, "Hull-House" (1898), 553; *Handbook of Settlements* (1911). In *JAP Microfilm*

(Continued)

Year	Investigation	Residents involved	Published as	Notes
1897	"Investigation of the Diet of the Italian Colony for the Department of Agriculture"	Caroline Hunt	W. Atwater and A. Bryant, *Dietary Studies in Chicago in 1895 and 1896* (1898).	Listed in: F. Kelley, "Hull-House" (1898), 553; J. Addams, "Jane Addams's Own Story ..." (1906); *Handbook of Settlements* (1911). In *JAP Microfilm*
1896–1898	"A Study of the Milk Supply"	Jane Addams	J. Addams and H. Grindley, "A Study of the Milk Supply of Chicago" (1898)	In *JAP Microfilm*
1896/1897	Recreation	George Hooker	G. Hooker, *A List of Pleasant Place for Nineteenth Warders to Go ...* [1896]	In *JAP Microfilm*
1897?	"General Study of the 19th Ward for Ethical Society of Chicago"			Listed in *Handbook of Settlements* (1911)
1897?	"Italians in Chicago" for U.S. Department of Labor	Caroline Hunt	C. Hunt, *The Italians in Chicago* (1897)	Listed in: F. Kelley, "Hull-House" (1898), 553
1900–1901	"Tenement House Investigation in Chicago"	Robert Hunter	R. Hunter, *Tenement Conditions in Chicago* (1901)	Listed in *Hull-House Bulletin* 7, 1 (1905–1906): 22; *Hull-House Year Book* 1906–1907, 54. In *JAP Microfilm*
1902	"Study of Typhoid"	Alice Hamilton, Maud Gernon, Gertrude Howe	Hull-House, *An Inquiry into the Causes of the Recent Epidemic of Typhoid Fever in Chicago* (1903); "An Inquiry Into the Part Played by the Housefly in the Recent Epidemic of Typhoid Fever" *The Commons* 8:82 (1903).	Listed in *Hull-House Bulletin* 5, 2 (1902): 14; *Hull-House Year Book* 1906–1907, 55; *Hull-House Yearbook* 1913, 49; *Hull-House Yearbook* 1916, 58. In *JAP Microfilm*
1902	"Study of Street Vending by Children"	Orville Wescott, Myron Adams (Buffalo NY)	*Newsboy Conditions in Chicago* (1907); M. Adams, "Children in American Street Trades" (1905)	Listed in *Hull-House Bulletin* 6, 1 (1903–1904): 16–7. In *JAP Microfilm*

Date	Study	Author(s)	Publication	Notes
1902	"Lodging House Investigation"	Orville Wescott	O. Wescott, "The Men of the Lodging Houses" (1903)	In *JAP Microfilm*
1903	"Study of Casual Labor on the Lakes"			Listed in: *Handbook of Settlements* (1911)
1903?	"Traction Study"	Henry Lloyd	H. Lloyd, *The Chicago Traction Question* [1903?]	In *JAP Microfilm*
1903–1904	"Investigation into the Selling of Cocaine"	Jessie Binford, Julia Lathrop, Alice Hamilton	"Profit in Child Victims to Cocaine" (1904)	Listed in: *Hull-House Yearbook* 1906–1907, 51; *Hull-House Yearbook* 1910, 51; *Handbook of Settlements* (1911); *Hull-House Yearbook* 1913, 51; *Hull-House Yearbook* 1916, 59. In *JAP Microfilm*
1905	"Study of Tuberculosis in Chicago"	Alice Hamilton, Theodore Sachs, Bertha Hazard	A. Hamilton, *A Study of Tuberculosis in Chicago* (1905); J. Addams and A. Hamilton, "The 'Piecework' System as a Factor in the Tuberculosis of Wage Workers" *International Congress on Tuberculosis Transactions* (1908).	Listed in: *Hull-House Bulletin* 6, 2 (1904): 20; *Hull-House Year Book* 1906–1907, 55; *Handbook of Settlements* (1911). In *JAP Microfilm*
1905	"An Intensive Study of the Causes of Truancy"	Gertrude Howe Britton	G. Britton, *An Intensive Study of the Causes of Truancy …* (1906)	Listed in *Hull-House Bulletin* 7, 1 (1905–1906): 22; *Hull-House Yearbook* 1906–1907, 49; *Handbook of Settlements* (1911). In *JAP Microfilm*
1905–1906	Re-investigation of Tenement House Conditions			Listed in: *Hull-House Bulletin* 7, 1 (1905–1906): 22.
1907	"Study of Midwifery" with Chicago Medical Society	Alice Hamilton	The Committee on Midwives, "The Midwives of Chicago" (1908); G. Abbott, "The Midwife in Chicago" (1915)	Listed in: *Hull-House Yearbook* 1910, 51; *Handbook of Settlements* (1911). In *JAP Microfilm*

(Continued)

Year	Investigation	Residents involved	Published as	Notes
1908	"Study of Greeks in Chicago"	Grace Abbott	G. Abbott, "A Study of Greeks in Chicago" (1909)	Listed in: *Hull-House Yearbook* 1910, 51; *Handbook of Settlements* (1911); *Hull-House Yearbook* 1913, 50. In *JAP Microfilm*
1909	"Study of Infantile Mortality among Selected Immigrant Groups"	Alice Hamilton	A. Hamilton, "Excessive Child-Bearing as a Factor in Infant Mortality" (1910)	Listed in: *Hull-House Yearbook* 1910, 51; *Handbook of Settlements* (1911); *Hull-House Yearbook* 1913, 50. In *JAP Microfilm*
1909?	Bakery conditions	Elsie Smith		Listed in: *Hull-House Yearbook* 1910, 54
1910	"Investigation of the Home Reading of Public School Children"	Gertrude Britton		Listed in: *Hull-House Yearbook* 1910, 51; *Handbook of Settlements* (1911); *Hull-House Yearbook* 1913, 51.
1910–1915	"Housing in Chicago, 1910–1915"	Sophinisba Breckenridge & Edith Abbott	S. Breckenridge and E. Abbott's series of articles in *The American Journal of Sociology* (1910–1911), parts I–V.	In *JAP Microfilm*
1925?	Mexicans and Health	Gertrude Howe Britton	"Our Mexican Patients in Central Free Dispensary"	Listed in: *Hull-House Yearbook* 1929, 64; *Hull-House Yearbook* 1934
1927?	"Gypsies in Chicago"		*Survey Graphic* (Oct. 1927)	Listed in: *Hull-House Year Book* 1929, 63
1929?	"Mexicans in Chicago"	Anita Jones	A. Jones, "Mexican Colonies in Chicago" *Social Service Review* 2, 4 (1928)	Listed in: *Hull-House Year Book* 1929, 63; *Hull-House Yearbook* 1934
1933	"Intelligence and Poverty Investigation"	Ethel Kawin	E. Kawin, "Intelligence and Poverty" (1933)	In *JAP Microfilm*

Sources: R. Woods and A. Kennedy, eds., *Handbook of Settlements* (New York: Charities Publication Committee, The Russell Sage Foundation, 1911), pp. 54; *Jane Addams Papers Microfilm* Reel 54 (note: the Microfilms do not merely list the investigations but are full copies of the published reports, some in book form).

Hull-House 1895, 55; Schultz 2007, 7).[8] In the various descriptions of *HHMP* by residents, I have not found a clear explanation of how the residents decided to create the maps. I believe they were well aware of Booth's maps and when Kelley was contracted to do her slum survey, they saw an opportunity to undertake a similar project.

Some scholars, such as van den Hoonaard (2013), give Kelley full credit for the maps. However, Agnes Holbrook provided the essay on "Map Notes and Comments" to *HHMP* describing how the data was collected and the maps prepared. Holbrook does not say the maps were hers, nor does she say Kelley is to be credited. She merely comments "a Hull-House resident" copied the schedules before they were sent to Washington and the maps were created from the data (Residents of Hull-House 1895, 55). In an essay on Hull-House published in 1898, Kelley appears to credit Holbrook for the maps as she discusses investigations conducted from Hull-House. She writes "Miss Agnes Holbrook's charted data from the investigations of the slums of great cities, made in the course of 1893 for the U.S. Department of Labor" and does not mention her own involvement in the maps (Kelley 1898, 553). Florence Kelley's son Nicholas remembers "residents" sitting in "the octagon," as Jane Addams office was called, and coloring the maps, with "Miss Holbrook" doing much of the coloring (Kelley 1954, 425). Artist Alice Kellogg Tyler captured Holbrook in a water-color sketch entitled "Miss Holbrook painting the Hull-House Maps," that was displayed at Hull-House as part of an art exhibition (Hull-House 1895).[9] My position is the maps were a collaborative effort of Kelley, Holbrook, and other residents, but Holbrook played a lead role in creating the maps. Agnes Holbrook's Wellesley College education included coursework in statistics and art, both of which prepared her well for translating geographical data into graphical forms such as maps (Dillon, undated).

There are eight maps in *HHMP*s – four of nationalities and four of wages – which can be assembled into two large maps covering the neighborhood south and east of Hull-House (Figure 4.1). Residents used the same color scheme used by Booth on his London poverty maps, but expanded from Booth's maps of income to include additional maps of ethnicity. Addams and Holbrook both acknowledge Booth's inspiration for their Hull-House maps (Residents of Hull-House 1895, viii and 11). While Booth depicted the whole of London by block-level, Hull-House covers a much smaller area, a neighborhood, and presents their data on the house or tenement-level, presenting an intimate portrait of a community. It also demonstrates wages and ethnicities can vary dramatically from house to house, and at times, within houses, rather than generalizing by block the income levels as Booth did. An 1895 review of the publication in *The Outlook* comments:

> A large part of the teaching of the volume will be conveyed by the maps setting forth by means of colors the nationalities and wages of the families in this district. These maps have the merit of simplicity. Any one by glancing at them may see just what the situation is.
>
> Books and Authors 1895, 785

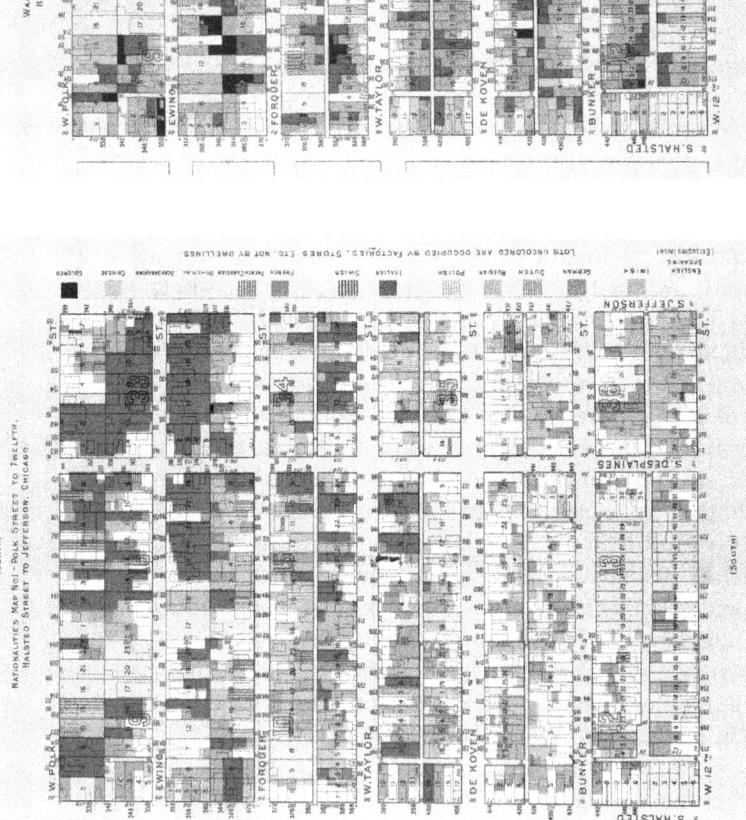

Figure 4.1 Details of the nationalities and wage maps from Hull-House Maps and Papers, "Map No. 1—Polk Street to Twelfth, Halsted Street to Jefferson, Chicago." North is at the top of the map. On the left, is the nationalities map. The darker colors towards the top of the map (blue) indicates Italians while the lighter shades (reds) towards the bottom of the map are Russians, Polish and Bohemians. White is used for "English Speaking (Excluding Irish)." On the right, is the wage map: "These figures represent the total earnings per week of a family." The darker colors (black and blue) indicate family wages of below $10.00 per week and the lightest (yellow) of over $20.00 per week. By permission of Special Collections and University Archives Department, University of Illinois at Chicago Library (HV4196C4H761895_c5_06 and HV4196C4H761895_c5_02).

The review discusses each map before turning briefly to a couple of the essays – Florence Kelley's on the sweating system and Eaton's on trade union employees. The review's focus on the maps is not unusual: the level of detail about nationalities and wages captured on these maps was unprecedented and they are widely recognized as benchmark work in social science research (Bulmer et al. 1991; Porter 2011, 294).

From Hull-House resident meeting minutes, it is clear that the maps were created with the intent to publish, with the note "What is to go with the maps?" indicating the essays were secondary to the maps (Knight 2005, 275). The maps were so central to the published work, they are named first in the title. Correspondence with the publisher reinforces their importance: when the publisher complained about the additional cost of publishing the maps in color and having them placed in a pocket in the back of the book, Addams and Kelley decided they would waive their royalties in exchange for the maps' publication (Deegan 1988a, 57–8).

In addition to the maps, there are ten essays: one on methodology by Holbrook, three on labor issues, three on ethnic groups in Chicago, and three on social work. While the title "Hull-House Maps and Papers" suggests a whole, the maps stand on their own, with essentially no connection to the essays on ethnicity. In fact, the ethnic colonies described in the essays are not included on the maps, although they are near Hull-House (Schultz 2007, 19). The studies, however, do include additional investigations conducted by Hull-House on labor conditions and "social studies" on particular ethnic groups.

The studies of labor conditions include work by Florence Kelley on the "The Sweating System" and Isabel Eaton on the conditions of the cloakmaking industry. Kelley's essay draws from her work at the State Inspector of Factories and Workshops for Illinois. As the Inspector, she published official reports detailing factory conditions and work hours and the women and children employed (Kelley 1894). The essay in *HHMP* allows Kelley a different medium to reflect on her observations than the official reports she published as well as allowing her to address, in particular, the conditions in the nineteenth ward, the Hull-House neighborhood. In the garment industry of the age, manufacturers depended on workers who were paid by the piece, working out of their homes or in small shops, using either hand-sewing or foot-powered machines, finding "leg-power and the sweater cheaper" than employing people directly and providing them with steam-power (Residents of Hull-House 1895, 70). In her work, Kelley found appalling working conditions, in crowded, poorly heated/cooled, poorly ventilated tenements, often infected with vermin and disease: "The unsanitary condition of many of these tenement houses, and the ignorance and abject poverty of the tenants, insure the maximum probability of disease; and diphtheria, scarlet fever, smallpox, typhoid, scabies, and worse forms of skin diseases, have been found in alarming proximity to garments of excellent quality in process of manufacture for leading firms" (Residents of Hull-House 1895, 66–7). Kelley places the blame on the manufacturers, seeing the workers as desperate to provide for their families and giving their children an education if they can. She calls on the abolishment by law of the sweating system, recognizing the growing presence of garment worker unions and their work towards this end, leading to the 1893 Workshop and Factories Act which made some headway in ending practices such as reducing child labor, separating homes from shops, and beginning an eight-hour work day for women and girls in Illinois (Residents of Hull-House 1895, 72). Kelley recognized this as but the start of the work that needed to be done to abolish tenement-house manufacturing. Kelley would continue to work on these issues, moving to the Henry Street settlement house in New York in 1899, and founding the National Consumers League.

Isabel Eaton's study is a comparative study of cloakmakers in Chicago and New York, using data provided by the unions and interviews with workingmen in the industry. Eaton found Chicago cloakmakers' incomes were slightly higher than those of New Yorkers, the hours of work comparable, with "every man is in debt" in both cities (Residents of Hull-House 1895, 94). The rooms of Chicago cloakmakers were generally better than those in New York, being slightly larger and slightly better in terms of light and ventilation, with Chicago paying lower percentage of their total income for rent. Yet this percentage was much higher than the "approved" percentage of income paid for rent, being approximately 29 percent in Chicago, 38 percent in New York compared to the "approved" 14 2/3 percent. She concludes "there is something very seriously wrong in the proportion of rent and wages in the cloakmaking trade" (Residents of Hull-House 1895, 95).

The "social studies" include an essay by Josefa Humpal-Zeman on Bohemians. Humpal-Zeman was an up-and-coming journalist and Hull-House resident who shared residents' interests in immigrants' experiences and women's rights (Noblitt and Zárasová 2001). In her essay, Humpal-Zeman begins by discussing the historic presence of Bohemians and their evolving spatial distribution in Chicago. By the 1890s, there were 60–70,000 Bohemians in Chicago, "the third largest city of Bohemians in the world" (Residents of Hull-House 1895, 107). Driven from Bohemia by wars, as well as social and political upheavals, skilled able-bodied men came to Chicago and established themselves as artisans, often investing in property through Bohemians building and loan associations (Residents of Hull-House 1895, 108). Humpal-Zeman then broadly addresses Bohemian life in Chicago, addressing labor issues, social life, family life, religion, and finally citizenship. She concludes by reflecting on the treatment of Bohemians and other immigrants by Americans and warns "... this constant aimless baiting of the American press gives these great masses one theme, one bond of sympathy, on which they can all unite; and that is,—hatred of Americans" (Residents of Hull-House 1895, 114). Yet, she suggests the Bohemians had not yet settled into hatred and, in fact, have "Americanize[d] almost too rapidly; so that frequently the second and third generations do not even speak their own native language."

The final essays focus on aspects of social work, with Julia Lathrop contributing an essay on the "Cook County Charities," Ellen Gates Starr contributing on "Art and Labor" (her passion), and finally Addams herself on "The Settlement as a Factor in the Labor Movement." Addams' essay essentially argues social settlements, in their efforts to improve living conditions and quality of life in their communities, have to support industrial organizations and enter the labor movement (Residents of Hull-House 1895, 140). Addams' approach, as is often true of her writings, is more philosophical, and offered broader reflections on American society based upon her extensive experiences at Hull-House. In many ways, she was the theorist for the community, as well as its figurehead. Addams writes of the challenges of organizing labor, where it involves bringing men as well as women together, often of a variety of ethnicities, as well as different political positions. Even something as simple as where to meet could be contentious, with a room over a saloon being acceptable with men but unacceptable to respectable women (Residents of Hull-House 1895, 141). She suggests not only can the social settlement provide a neutral setting for meetings but it can assist the labor movement in focusing on the "larger and steadier view" to advance democracy, to not react only to "wrongs." Addams argues the growing "fund of altruism" in American society has led to

> "… the motive force which is slowing enfranchising all classes and gradually insisting upon equality of condition and opportunity. If we can accept this explanation of the social and political movements of our time, then it is clear that the labor movement is at the bottom an ethical movement, and a manifestation of the orderly development of the race.
>
> Residents of Hull-House 1895, 148

Addams would continue to write, speak, and advocate for the interplay of democracy and ethics for the rest of her life. Hull-House and its work was ultimately an expression of Addams' views of what it meant to live in a democratic society with citizens afforded rights as well as responsibilities to other citizens.

Hull-House Maps and Papers were very well received but Hull-House would not publish another stand-alone volume on their investigations. Rather, the various Hull-House investigations were published through a variety of means – some as pamphlets, some as articles in leading journals such as the *American Journal of Sociology*, *The Survey*, and the *Journal of the American Medical Association*, and some as books focused on particular issues, such as Robert Hunter's *Tenement Conditions in Chicago* (1901) or Sophinisba Breckenridge and Edith Abbott's *The Delinquent Child and the Home* (1912). They often published the results in multiple mediums, expanding the audience for their work. For example: residents' work on typhoid was published as a pamphlet, as an article in *The Commons*, and as an article in the *Journal of the American Medical Association*

(Gernon et al. 1903a; Hamilton 1903; Hull-House Residents 1903). The investigations conducted by Hull-House covered an amazing breadth of topics germane to life on Chicago's South Side, including work on epidemics, "social studies" on the various ethnic enclaves found in Chicago, on the living conditions for various groups, and on issues they felt strongly in need of attention, particularly regarding the treatment of children (child labor conditions, problems with cocaine use, etc.).

Hull-House residents and physician Alice Hamilton led residents in studying and mapping two Chicago epidemics, one of typhoid in 1902 and tuberculosis in 1905 (Hull-House 1903, 1905). In the studies, it is clear the contributions of each from their title pages: "House-to-House Investigations by Miss Maud Gernon and Miss Gertrude Howe. Bacteriological Inquiry as to the Relation of the Common Housefly to the Spread of the Epidemic, Dr. Alice Hamilton, Research Conducted at Memorial Institute for Infectious Diseases." The analysis of the typhoid epidemic investigation by Hamilton concludes sanitation conditions do not appear to be the only cause of typhoid and hypothesize flies may be a more influential factor. The maps capture the outbreaks at a variety of scales (Figure 4.2). The typhoid maps depict deaths from typhoid on a city-wide basis and then in a small west side area severely affected. The maps are integral to the text, being referred to

Figure 4.2 Detail from Gernon, Howe and Hamilton's *An Inquiry into the Causes of the Recent Epidemic of Typhoid Fever in Chicago* (1903a). The map is of a small area on the West Side within the Nineteenth Ward where there was a high proportion of the deaths. The shading indicates the state of the sanitation conditions, from flushing toilets down to outhouses. The dots indicate the locations of typhoid deaths. Data was collected from house-to-house investigations of 2,002 dwellings in the area.

numerous times: such as "The accompanying map shows "..." (p. 5), "It will be seen from the maps "..." (p. 6), and "It is probable that the number of cases, as given in the maps...." (p. 8). Hamilton's article in the *Journal of the American Medical Association* reprinted the maps.

The tuberculosis study was conducted by Hamilton, Dr. Theodore Sachs, and Bertha Hazard, a Hull-House resident; a house-to-house investigation of tuberculosis in the Jewish district of Chicago. They found a lower mortality rate among Jewish residents than non-Jewish residents in the district and, overall, appalling housing conditions with congested homes with damp rooms and no ventilation (Hamilton 1905, 6). The tuberculosis maps produced from their investigation capture the outbreak on a variety of scales, from the district level to a single block in the Jewish district, depicting the deaths from tuberculosis, as well as listing the population of the block and breaking it down further into Jews/non-Jews, adults/children. Hamilton, in conclusion, called for improved treatments to mitigate against tuberculosis, the tracking of diagnosed tuberculosis cases, improved diet for those afflicted, and the disinfection and renovation of infected homes. As one of the nation's first specialists in occupational medicine and health, Hamilton eventually took a position at Harvard Medical School but essentially split her time between Harvard and Hull-House.

Hull-House residents continued their "social study" investigations, with many of the studies published in the *American Journal of Sociology*. Edith Abbott and Sophonisba Breckenridge, both residents, were eventually faculty at the University of Chicago. Between 1911 and 1912, Abbott and Breckenridge published five articles on Chicago housing conditions which included maps. Further studies on ethnic enclaves, such as Alzada Comstock's "The Problem of the Negro" and Natalie Walker's "Greeks and Italians in the Neighborhood of Hull-House," included maps, with Comstock illustrating the "Black Belt in Chicago" and Walker illustrating the density of residents in the neighborhood of Hull-House. Abbott and Breckenridge's third article, 'Back of the Yards' (1911a), is indicative of the Hull-House approach to neighborhood studies, involving extensive surveys and interviews, maps and statistical analyses (Figure 4.3). The 'Back of the Yards' refers to the area between the Union Stockyards ('Yards') and the city dump. At its worst, the neighborhood featured unpaved streets, no sidewalks, accumulated garbage, and no effective real sanitation, and Abbott and Breckenridge's work occurred after some improvements had been made to streets and sidewalks and to sanitation. Yet the area still had high rates of infant mortality and tuberculosis, large numbers of applicants for poor relief, overcrowded housing conditions, and large numbers of immigrants. As authors, they wondered "what the actual living conditions might be in a dismal region" (Breckenridge and Abbott 1911a, 436–7). House-to-house canvassing was done, sampling Polish and Lithuanian blocks and an area known for its high proportion of lodgers in an attempt to capture 'typical homes of this neighborhood' (Breckenridge and Abbott 1911a, 437). They presented

a)

b)

Figure 4.3 Example of a "social studies" map: "Map showing the district between the Union Stockyards and the City Dumps" from Breckenridge and Abbotts' "Housing Conditions in Chicago III: Back of the Yards" (1911).

their evidence in the forms of data tables, a map, and photographs, covering nationalities, occupations, congestion by block and by room, sanitation (privy in yard or in building), owning versus renting, cost of rent, and number of saloons. Through their data, they presented empirically-derived facts as they encountered the conditions directly in this neighborhood. To the unobservant eye, there appeared to be space in the form of vacant lots and free-standing two-story buildings, but this was deceptive: the buildings were extremely congested and living conditions appalling. In their conclusion, Abbott and Breckenridge call for the enforcement of existing tenement laws and the expansion of the city's sanitation department to enforce such laws (Breckenridge and Abbott 1911a, 468). Abbott and Breckenridge continued to research and publish together throughout their lives in addition to their independent work (for example, see Breckenridge and Abbott 1912, 1936).

In a similar study to 'Behind the Yards,' Alzada Comstock examined the living conditions of Black Chicagoans (Comstock 1912). Unlike the 'Yards,' Comstock's research was not concentrated on an area but on a population distributed over four districts, the largest of which she referred to as the "black belt" (Comstock 1912, 242). Like the "Yards," the location of housing was largely tied to occupation, thus the "black belt" was located near the railroads, the largest employer of Black Chicagoans. House-to-house canvassing was used to examine two of the largest districts, focusing on race/ethnicity, families versus lodgers, housing conditions, sanitation, and housing congestion. Comstock found only the "black belt" was predominantly black, the second largest area was only one-third black. Comstock's major finding was that "the rent paid by Negroes is appreciably higher than that paid by people of any other nationality" (Comstock 1912, 253). Housing was typically in poor repair, so "the immigrant, for a smaller amount of money, may live in a better house than the Negro" (Comstock 1912, 255). She rightly concluded prejudice was the root of the problem. Unlike Hamilton, and Abbott and Breckenridge, Comstock did not offer remedial steps to improve the situation: perhaps, then as now, there is no simple solution to solve pervasive societal problems such as prejudice.

In addition to published studies, there were a number of unpublished studies conducted by Hull-House. Alex Elson, a child of Russian immigrants, grew up at Hull-House. In the 1950s, he recalled:

> The Hull-House neighborhood became an area of intensive study. Nationality distribution, occupation and income level, size of families, kind and character of housing, were reduced to charts and maps, some of which were put on the walls of Hull-House so that the neighbors could have the benefit of this research.
>
> Elson 1954, 6

Elson's memory of Hull-House with its maps on the walls suggests a place where spatial knowledge was created and circulated beyond the published maps. Reviewing *Hull-House Bulletins*, a periodic newsletter of Hull-House activities, has uncovered references to such maps as well as additional investigations:

> During the last year several investigations have been made from Hull-House, one concerning the nationalities to be found in one-third of a square mile directly southeast of the house. A nationality map of this district was made in 1894, and the changes brought about during the eleven years which have elapsed since then are most strikingly portrayed by comparing the new map with the old.
>
> Hull-House 1905–1906, 22

To my knowledge, a nationalities map of the Hull-House community has not been published since the original *HHMP* of 1895; it may have been created for their use only.[10] With references to studies in *Hull-House Bulletins*, in articles published on Hull-House and their work, and in individual's accounts, we are aware more investigations and mapping took place than was published, such as the "tramps investigation," "General Study of the 19th Ward for the Ethical Society of Chicago," "Study of Casual Labor on the Lakes," and "Reinvestigation of Tenement House Conditions" (Table 3.1).[11] Moreover, this work was made available at least to the residents, if not to the neighborhood citizens. They were producing and sharing their knowledge of the neighborhood and fostering to some extent a knowledge community, to use a contemporary term. Scholar Mary Jo Deegan has commented on the residents' use of cartography:

> Not only was this methodological technique first used in Chicago with HH Maps and Papers, but Hull-House residents openly continued this tradition and practice. Any visitor to Hull-House, let alone a scholarly one, was immediately exposed to their enchantment with mapping.
>
> Deegan 1988a, 46[12]

Deegan suggests by conducting such studies and creating maps: "the maps of Hull-House were intended to reveal to the people of the neighborhood that their lifestyles had patterns and implications that they could use to make more informed decisions. These maps were part of the community, and integral to the settlement's goals of democracy and education" (Deegan 1988a, 47).

The practice of geography and mapping by the residents of Hull-House uniquely linked social science with social activism, as demonstrated by their community studies (Schultz and Hast 2001, xxxiv). Holbrook, in her essay on the Hull-House maps, wrote: "Merely to state symptoms and go farther would be idle; but to state symptoms in order to ascertain the nature of disease, and apply, it may be, its cure, is not only scientific, but in the highest sense humanitarian" (Residents of Hull-House 1895, 14). Addams describes, in an 1899 paper on the functions of social settlements, a settlement being approached by a "head of a federal department" and begging them:

> ... to transform into readable matter a certain mass of material which had been carefully collected into tables and statistics. He hoped to make a connection between the information concerning diet and sanitary conditions, and the tenement house people who sadly needed this information.
>
> Addams 1899, 338

Addams sagely observed the federal department head missed the point: "to put information into readable form is not nearly enough. It is to confuse a simple statement of knowledge with its application" (Addams 1899, 338).

Practicing geography and cartography at Hull-House and other settlement houses

> There are classes, it is true, but receiving and imparting information are not their only ends – study is almost an outgrowth of social life.
>
> Holbrook 1894, 172

Writing about the atmosphere of inquiry that permeated Hull-House, Agnes Holbrook observed investigations and education were not just part of the work of Hull-House but "an outgrowth of social life," an aspect of Hull-House culture. Educational programs were created to serve community needs, such as its kindergarten and English classes, and space was made for programs such as the Working People's Social Science Club and Hull-House Women's Club. Informal opportunities for engagement with new ideas were present throughout Hull-House: over dinner, often with distinguished guests like Kropotkin or Stead, presentations, debate and argumentation in various clubs, such as the Working People's Social Science Club, and, eventually, classes in the Chicago School of Civics and Philanthropy (Jackson 2000, 165). In addition to the educational programs for the neighborhood, Hull-House offered higher educational opportunities for its residents, workers, and others. Its programs predate the extension courses offered by universities in Chicago and eventually became one of the early university extension centers (Fish 1985, 39; Holbrook 1894, 174; Residents of Hull-House 1895, 153). For all those present in Hull-House, discussion after hours would continue in the parlor or in residents' rooms (Abbott 1950, 377). Beatrice Webb would write of her visit in 1898: "One continuous intellectual ferment is the impression left on the visitor to Hull-House" (Webb 1963, 108).

Hull-House functioned as a graduate school of sorts in social work with residents finding interests which they pursued lifelong (Fish 1985, 189; MacLean and Williams 2012, 247). The studies they participated in often involved household visits and social surveys, data analysis, the creation of maps and graphs, the formulation of proposals for social change, and they made the results available through a variety of mediums (Lengermann and Niebrugge-Brantley 2002, 102–3). In time, such pursuits and training led to the creation of the Chicago School of Civics and Philanthropy (CSCP), "a community school established at Hull-House in 1903 to educate and train workers for social service and to conduct research on social problems stemming from industrialization" (MacLean and Williams 2012, 239). The origins of the Chicago School of Civics and Philanthropy are complicated: its earliest permutations date from 1895. Hull-House residents such as Julia Lathrop, Jane Addams, and Mary McDowell were involved with its creation and establishment (Coohey 1999). Abbott and Breckenridge taught "Methods of Social Investigation" at the CSCP:

> A course designed to acquaint students with the most important work done in the field of social inquiry. Some reports of Royal Commissions and of the English and American Labor Departments, and other selected official reports, together with such private investigations as Booth's Life and Labour of the People of London, Rowntree's Poverty, Women in the Printing Trades, the Dundee and West Ham inquiries and the Pittsburgh Survey will be discussed. The application of statistical methods to social problems, the collection and tabulation of data, the use and misuse of averages, index numbers and weighting will be treated briefly; the use and limitations of the experiment, the interview, the document and personal observations will be considered.
>
> Chicago School of Civics and Philanthropy 1911, 190

The CSCP was transferred to the University of Chicago in 1920 and renamed the School of Social Service Administration. Individuals such as Edith Abbott and Sophonisba Breckenridge would become university faculty members, taking their Hull-House perspective into the universities, albeit often in "women's departments" such as household administration, sanitary science, social settlements and statistics (Deegan 1981, 15).

As social settlements grew across the country, books, publications and education programs promoted such investigations and described how to conduct them. Charles Henderson, who taught sociology at the University of Chicago, published a book on social settlements in 1899. Henderson's volume covers the history of their development, overviews of settlement houses in the United States, the theories underlying their practice, and then their actual implementation. As part of Henderson's discussion of "Laying Foundations," selecting a neighborhood to work in, he advocates mapping:

> In order to make a wise choice of fields, a preliminary survey should be made. A large map of the district should be drawn, and on it should be set down the essential social facts of significance, dwellings, population,

saloons, schools, industries, nationalities represented, churches, missions, thoroughfares, lines of travel, places of amusement, sanitary conditions, water supply, branches of libraries, fire departments, police stations, etc.

Henderson 1899, 107

Similarly, in 1926, Mary Simkhovitch published *The Settlement Primer.* Simkhovitch, "Headworker" of the Greenwich settlement house in New York City, provides an overview of settlement work, from establishment through types of activities, finances and "What to Avoid."[13] From the very beginning, Simkhovitch states "The settlement should know its own neighborhood, its housing, health, recreation, industries, family and social life, political and religious associations" (Simkhovitch 1926, 9). She goes on to emphasize the importance of neighborhood understanding, through becoming part of it as well as through surveys: "No year should be permitted to pass without some tangible and recorded addition, no matter how humble, in the field of social research. Of course, no House should be without a descriptive map of its district, a directory of its principle features" (Simkhovitch 1926, 11).

By the early twentieth century, practices of social survey work and mapping were being taught as an aspect of sociology and social work education. Emily Greene Balch, who taught at Wellesley, used mapping in both her teaching and her research:

We hear of Miss Balch's students in 1901–2 mapping the North End of Boston.... Each student handed in a final paper describing the section from personal observation, together with a 'social map' showing centres of social significance – schools, hospitals, saloons, churches and settlement houses.

quoted in Platt 1994, 60

A 1928 student manual for *Field Studies in Sociology* by Vivien Palmer has a whole chapter on "The Social Research Map." Palmer, who was coordinator of the urban research program in the Sociology Department of the University of Chicago, emphasizes the importance of maps to not only display findings but also as "... part of his working technique for locating his problem and for analyzing and discovering relationships in data" (Bulmer 1991a, 304; Palmer 1928, 185). Palmer goes on to discuss the various ways maps can be used (distributions, correlations, movements and trends), the importance of the base map, basics of map making and finally map interpretation.[14] Palmer however differentiated between social surveys and sociological surveys, "emphasizing that the latter was not concerned with reform and amelioration but with 'the scientific discovery of how human societies function'" (Bulmer 1996, 28).

As social surveys and mapping became codified and as sociology and social work were established as academic disciplines, Hull-House led the charge in their practices. Agnes Holbrook in her essay "Map Notes and Comments" in *HHMP* describes their maps "as an illustration of a method of research." While both Holbrook and Addams are quick to describe their work as a "snapshot" or "recorded observations," Holbrook does go on to state "Hull-House offers these facts more with the hope of stimulating inquiry and action, and evolving new thoughts and methods" (Residents of Hull-House 1895, 45 and 58).

Hull-House was indeed the innovator and other settlement houses followed their example, in both actions and in publications. While the Hull-House maps had been inspired by Charles Booth's, the Hull-House maps in turn, inspired other mapping projects, such as W.E.B. Du Bois' *The Philadelphia Negro* (1899) and *The Pittsburgh Survey* (1909–1914), which all employed social surveys and mapping as well as other methods of presenting data such as charts, graphs, tables, and photographs (Lengermann and Niebrugge-Brantley 2002, 11).

W.E.B. Du Bois was asked by the University of Pennsylvania and the College Settlement of Philadelphia to conduct an in-depth survey of the Black community in Philadelphia, focusing on its Old Seventh Ward, where 10,000 of Philadelphia's 40,000 Blacks lived. The concept of the Philadelphia study originated with Susan Wharton, whose family founded the Wharton School of Economy and Finance at the University of Pennsylvania, and who was a member of the College Settlement of Philadelphia's executive committee (Deegan 1988b, 305). Du Bois had a Ph.D. in Social Science from Harvard and was teaching at Wilberforce University in Ohio. Du Bois would move his family into the neighborhood in 1896 and conduct extensive interviews, supplemented by archival research and census data (McGrail 2013; O'Connor 2009). He eventually produced *The Philadelphia Negro* in 1899 in an effort to "ascertain something of the geographical distribution of this race, their occupations and daily life, their homes, their organizations and, above all, their relation to their million white fellow-citizens" (Du Bois 1899, 1). But Du Bois placed less emphasis on social problems, intending his study to be "a work of social science as well as an examination of social conditions … and to show that black Philadelphians were a product of their social environment" (Bulmer 1991b, 175). In addition to detailed statistics on the community with tables and charts, Du Bois created a striking map of the ward color-coded to their social class (Hillier 2009, 40). Like Hull-House maps, Du Bois depicts his data on a household basis.

Working with Du Bois was Isabel Eaton. Eaton had lived and worked at Hull-House and contributed to *Hull-House Maps and Papers*. Wharton corresponded with Addams regarding the planning of the Philadelphia study: it is assumed Addams recommended Eaton for the project (Deegan 1988b). In her work with Du Bois, Eaton focused on interviewing domestic workers and produced a report on domestic service that is an appendix in *The*

Philadelphia Negro. In some editions, she is listed as a co-author with Du Bois (Schultz 2007, 37). Her study of domestic workers would earn Eaton a Master's degree from Columbia University. I had initially missed Eaton's contribution to the volume and assumed Du Bois was the sole author. Evidently this is a common error: "Despite her substantial contribution, none of the analyses of *The Philadelphia Negro* written by men acknowledge her substantial input" (Deegan 1988b, 307). But her study on domestic workers, largely women, was benchmark work in its own right and often cited (see, for example: Katzman 1981, 306; Schultz 2013; Vapnek 2009, 107).

The Pittsburgh Survey project was deeply tied to the magazine *Charities and the Commons,* the leading American magazine on reform. Editor Paul Kellogg had been asked by the chief probation officer of the Juvenile Court of Alleghany County to investigate the conditions in Pittsburgh and surrounding communities. With funding from the Russell Sage Foundation, seventy-four "investigators" went into the field, exploring different aspects of "intolerable urban conditions" (Cohen 1991, 247–9). Initially, thirty-five articles were published in *Charities and the Commons.* Eventually, the articles and additional material were compiled into the six-volume *Pittsburgh Survey* (Kellogg 1909-1914). Some of the volumes are compilations of the articles, others are freestanding studies, such as volume 3, Elizabeth Butler's *Women and the Trades* (1909). Butler's study examined the experiences of 22,000 women working in 448 shops and factories in the Pittsburgh area, interviewing managers as well as workers (Butler 1909, 380–1). She found women working "… in factories that produced pickles, confections, crackers and cake, cheap cigars, workers' overalls, paper boxes, clothbound books, glassware, and electrical equipment" (Greenwald 1996, 147). Butler's extensive interviews and statistics were illustrated with maps locating industries where women worked and with photographs by Lewis Hine. A map of "Locations of Establishments Employing Women" shows there to be a dense cluster in the "Three Rivers" area but generally establishments could be found throughout the city. Another map shows the location of "Stogie factories and sweatshops," accompanied by a graphic depicting how different branches of the stogie industry were gendered: 100% of the drying room was men while the teachers, stripping, and packing were almost entirely women. While it might be assumed women's employment is less dangerous than men's, Butler found that many women's occupations were dangerous to women's health, especially in industries such as lacquering, tobacco stripping and steam laundries (Kleinberg 1996, 98). Butler also found the income women earned was important to their families, with many being the chief breadwinners (Butler 1909, 348). But despite the import of this income, women faced many inequalities. In the introduction to *Women and the Trades*, Kellogg recounts the story of a well-trained young man giving a tour of a factory. He informed the guests the hundreds of girls working at high-speed machines received $20 per week, a fact the guests found hard to believe. The young man was confident and called for the payroll department to support his claim: Only the forewoman received $20, the remainder received $5–$7 a week. Despite the importance of their labor to their families, women's jobs were generally low paying and second-class. Like Booth's London maps, the *Pittsburgh Survey* was displayed and circulated, such as at Harvard, as described by historian Frederick Jackson Turner, who recalls:

> … an exhibit on the work in Pittsburgh steel mills, and of the congested tenements. Its charts and diagrams tell of the long hours of work, the death rate, the relation of typhoid to the slums, the gathering of the poor of all Southeastern Europe to make a civilization at that center of American industrial energy and vast capital that is a social tragedy.
>
> Turner 1914, 249

The Philadelphia Negro and *The Pittsburgh Survey* are examples of large, extensive studies but many of the social settlements produced smaller studies, often published as brochures or as articles in journals. In publications such as the *Handbook of Settlements* (1911) and *Bibliography of College, Social, University and Church Settlements* (1905), a short description of the settlement is followed by a list of publications, some describing the settlement and others investigations by the settlement. Through these sources, it is possible to get a glimpse of the range of investigations conducted out of social settlements. For example: the Neighborhood House, another Chicago cooperative neighborhood social center, incorporated in 1903, grew from a kindergarten established by Hull-House friend Harriet Van Der Vaart. In addition to articles about their work published in a variety of publications, Van Der Vaart published investigations on child labor (Van Der Vaart 1902, 1903, 190x).

The University of Chicago Settlement was founded in 1894 and, while affiliated with the University, it was not funded by it (Wade 2001, 563). Its head resident was Mary McDowell, a friend of Hull-House and another socially active Chicago woman. The Settlement was located west of the Stockyards or what they called "Back of the Stockyards." In the 1905 *Bibliography of College, Social, University and Church Settlements*, eight "social studies" were reported published by McDowell and a resident between 1900 and 1905, covering topics such as women workers, women labor unions, and Bohemians in Chicago (Montgomery 1905, 36–8). In 1913, Louise Montgomery of the Chicago Settlement published *The American Girl in the Stockyards District*, an in-depth study of girls raised in the Stockyards District, examining their access to education, their experiences in the work-force, and problems of adjustment (Montgomery 1913, 2).

The social studies or surveys conducted by Hull-House and other settlement houses predate the first sociology departments and foundations, "… in a sense we were actually pioneers in field research," wrote Addams in 1930 (Addams 1930, 405). This field-based work places an emphasis on being rooted in the community, something Addams and residents took pride in. As one Hull-House neighbor explained, Hull-House was "well grounded in the mud. This was at once a statement and a compliment" (Moore 1897a, 630). Their claims of situated knowledge were used to emphasize their knowledge of the neighborhood was based on first-hand experience, time spent in the community, and upon in-depth surveys.

"Ocular proof": Mapping by affiliated women's organizations

I wear it [the suffrage map] sandwich fashion, and walk about my crowded streets. It attracts everyone's eye, and an explanation of the colors excites deep interest and makes a great impression. Men are much impressed by the ocular proof of our advance, and after little talks in groups of three to ten, many sign slips. The colored map is, I think, very valuable, as many people receive impressions more strongly through the eye than the ear.

Walks and Wins ... (1913).

While I cannot definitively say Hull-House's "enchantment with mapping" led directly to the use of mapping by other Progressive Era women's organizations, I can say associations housed at Hull-House, such as the Immigrants' Protective League and the Juvenile Protective Association, conducted social studies and included maps and mapping in their publications (Schultz and Hast 2001, xxxv). The "ocular proof" offered by mapping proved to be a powerful persuasive instrument in their various campaigns.

Hull-House residents and associates were active in a number of organizations that conducted social studies and created and used maps, such as the Women's Christian Temperance Union, Young Women's Christian Association, Chicago Women's Club, Women's Trade Union League, and city, state, and national suffrage organization. Hull-House was an important hub for Progressive Era activism, especially among women. It was not unusual for Hull-House residents or affiliates to be involved in a number of organizations. For example, Sophonisba Breckenridge, author of many of the ethnographic studies, a one-time resident of Hull-House, a member of the University of Chicago's faculty, was active with the National American Suffrage Association, Women's Trade Union League, the Women's City Club of Chicago, the Woman's Peace Party, the Immigrants Protective League, and the National Association for the Advancement of Colored People (Fish 2001, 115). Florence Kelley, writing in 1898, observed:

One key to the growth of the House is probably the fact that many residents have taken active part in those municipal, social and labor movements which bear indirectly as well as directly upon the life of the neighborhood. The Arbitration Congress, the Child Labor law, the garment workers' strike, the Charity Organization Society, the Civic Federation, the

Chicago Women's Club and many other elements of the life of the city and state have made heavy demands upon the energies of the House.

Kelley 1898, 558

As women participated in multiple groups, they could theoretically keep the groups informed of each others' activities, possibly coordinate their efforts, and potentially transfer successful methods and tactics from one group to another. Moreover, the groups often collaborated and supported each other: Women's clubs of the settlement houses sold buttons with "Votes for Women" on one side and "Women's Trade Union League" on the other (Flanagan 2005). It is not possible to trace out the complete network of women's activism in Chicago, but I will highlight some of the major efforts: suffrage, National Women's Trade Union League, Immigrant's Protective League, Woman's Christian Temperance Union, and women's legal rights advocacy.

The American suffrage movement and its Illinois chapter had numerous ties to Hull-House with Jane Addams, Florence Kelley, Sophinisba Breckenridge, Mary McDowell, Grace Abbot and other residents all working at various times with Chicago, Illinois, and national suffrage organizations. Addams published numerous essays in support of suffrage, including "Why Women Should Vote" in *Ladies' Home Journal* in 1910, and was elected vice-president of the National American Woman Suffrage Association, serving from 1911 to 1914 (Brown 1995, 183 and 185). A major effort to try to obtain municipal suffrage in Chicago was launched in 1906 when Chicago attempted to write a new municipal charter. Women's groups organized to demand municipal suffrage be added to the new charter. When their proposed charter failed, "over a hundred women's organizations, including working-class and immigrant women, waged a successful campaign to urge male voters to defeat the charter when it was put before Chicago voters" (Flanagan 2005). As part of the campaign for municipal suffrage in Chicago in 1906, a "suffrage map" was created by Anna Nicholes, a Hull-House resident and active member of the National Women's Trade Union League (McCulloch 1908). Unfortunately, I have not been able to locate a copy of this map although there are enough references to it to establish it did indeed exist. This prototype suffrage map created by Nicholes led to a national suffrage map used extensively in the campaign for suffrage. See Chapter 5 for an extensive discussion of the national suffrage map. Chicago women received municipal suffrage in 1913, and nationally suffrage in 1919 (Wade 2001, 564).

The Illinois branch of the National Women's Trade Union League was housed at Hull-House from its inception in 1904 until 1908. Founded when the American Federation of Labor would not allow women to join, their objective was "to investigate the conditions of working women, to promote the best type of trade unionism in existing organizations and to assist in organizing trade unions among women." (Kirkby 1992, 36). Their members included some thirty-two different labor groups including: Associated

Vaudeville Artists, Broom Makers, Commercial Telegraphers, Italian Garment Workers, Necktie Workers, Teachers' Federation, Waitress' Union. They fought for such radical rights as "an eight-hour workday, a minimum wage, an end of women's night work, and the abolition of child labor" (Jackson 2005, 715). They were affiliated with most of the women's clubs and organizations in Chicago. Jane Addams was on their board of directors, Sophonisba Breckenridge was chairman of their Investigation Committee, and Anna Nicholes, chairman of their Benefit Committee. The Benefit Committee, which provided assistance to workers unable to work due to illness, met regularly at Hull-House. They also had a library, a chorus (which met at Hull-House), social gatherings where they hosted dignitaries from other countries (such as a member of the English Parliamentary Labor Party and English suffrage leaders), and two national publications. At "regular meetings," guest lecturers covered topics such as "English Women and the Vote," "The Eight-Hour Day and Legislative Protection of Women," and "Tuberculosis in Its Relation to Industrial Conditions." Representatives from the WTUL spoke to a wide range of union meetings as they in turn represented over thirty organizations. During the 1910–11 garment workers' strike, they provided food aid to strikers and their families and helped draft an agreement ending the strike at a major Chicago manufacturer (Higbie 2005).

In the 1905–1906 *Hull-House Bulletin*, the WTUL reported they carried out a number of investigations – on the conditions of clerks over holidays, of the conditions of women working in certain trades, and of the laws by state protecting women in employment (*Hull-House Bulletin* 1905–1906, 21). After the Triangle Shirtwaist Factory Fire in 1911, NWTUL conducted a four-year investigation of factory conditions which led to new health and safety regulations being established (Jackson 2005, 716). I have not been able to locate any of these studies to date and so cannot comment on their geography or maps. But in a 1915 brochure on "The Eight-Hour Day for Women," they published a map of states limiting hours of women's work as part of their advocating for an eight-hour work day for women, with the states who successfully passed eight hour work day laws in white and states with no limitations in black (Figure 4.4). A 1917 article in the NWTUL magazine *Life and Labor* on the campaign for the eight-hour work day in Nevada is entitled "A Black Spot Off the Map," despite the article not including a map (Andrews 1917). While the metaphor of "black spots" on maps had been part of Western culture for at least 100 years at that point, the phrase was used extensively in the suffrage campaign as well as in the temperance movement (see Chapter 5).

From the work of the Illinois chapter of the NWTUL, came the creation of the Immigrants' Protective League in 1908 (initially called the League for the Protection of Immigrants). Immigrant women and girls were particularly vulnerable, especially traveling alone from their home countries to the United States. In her foreword to the 1909–1910 Annual Report of the Illinois chapter, Addams wrote:

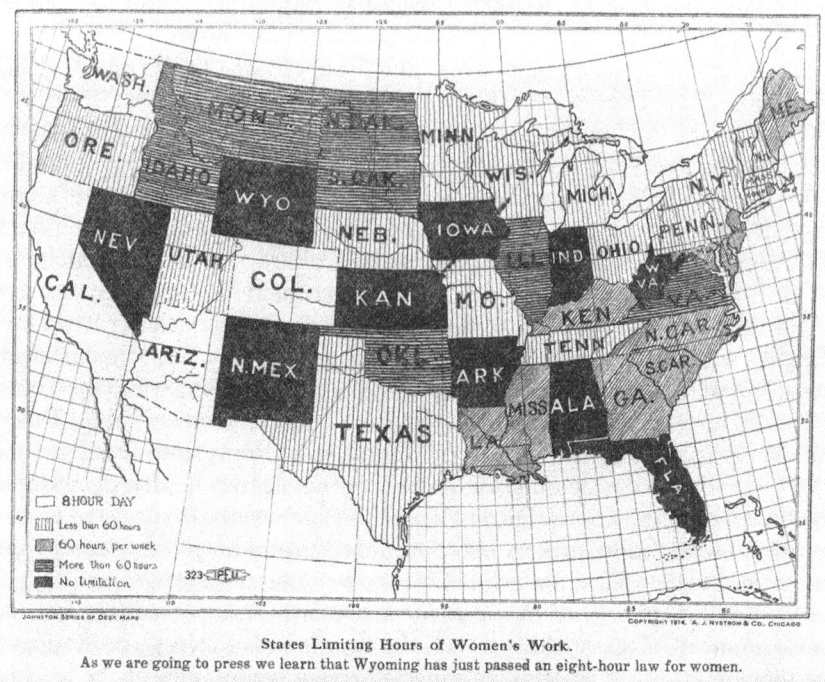

States Limiting Hours of Women's Work.
As we are going to press we learn that Wyoming has just passed an eight-hour law for women.

Figure 4.4 National Women's Trade Union League map from the pamphlet "The Eight-Hour Day for Women" published in 1915. States with an eight-hour day are in white, the striped are less than 60 hours, the medium shade is 60 hours per week, the second darkest shade is more than 60 hours and black is no limitation. By permission of the Wisconsin Historical Society, Madison, WI (Pamphlet 58-3035; Image ID 130591).

Every year we have heard of girls who did not arrive when their families expected them, and although their parents frantically met one train after another, the ultimate fate of the girls could never be discovered; we have constantly seen the exploitation of the newly arrived immigrant by his shrewd countrymen in league with the unscrupulous American; from time to time we have known children detained in New York and even deported, whose parents had no clear understanding of the difficulty.

League for the Protection of Immigrants 1910, 4

In 1920, it was estimated 20% of women and girls leaving Ellis Island for Chicago did not reach their destination (Buroker 1971, 646). The League sought to reach out to newly arrived immigrants, assist them finding housing and legitimate work, and introduce them to resources available to them, such as English classes.

Besides Addams, Breckenridge and Grace Abbott were active with the League. Towards their goals, they "collected information of Illinois immigrant groups, established a case work service, aided new arrivals, improved the employment agency situation, and pressed for state and federal legislation" (Buroker 1971, 654). They located missing immigrant girls. And they conducted studies to advance their work, such as studies of "new" immigrant groups such as Bulgarians and Greeks, a study of Chicago employment agencies, a study of the educational needs of Illinois immigrants, and a study of immigrants and the coal mining communities of Illinois (Buroker 1971, 656; League for the Protection of Immigrants 1910, 6). The work of Abbott and others was so recognized, "In 1914 Abbott and some other Chicago social workers were invited by the state of Massachusetts to conduct a statewide survey of immigrant problems" (Buroker 1971, 657).

Hull-House residents also had ties to the Women's Christian Temperance Union. Mary McDowell was a Hull-House resident, a national organizer for the WCTU, and eventually headed the University of Chicago Settlement House in Chicago. A photograph of the Hull-House residents' dining hall dating from early twentieth century depicts WCTU materials on display on its walls (Glowacki and Hendry 2004, 69). WCTU was founded in 1873; its most prominent member and long-serving president was Frances Willard.[15] It was "the largest and most influential activist movement of women in the country ... taking an active role and powerful role in the passage of the 18th and 19th amendments to the U.S. Constitution, as well as hundreds of other laws affecting women and children" (Mattingly 2008, 133). According to Mattingly,

> Women protested the manufacture and sale of alcohol, which they believed contributed to women's and children's hardships, but for many the focus expanded to include broader women's issues, especially unequal treatment under the law. ... Because alcoholic men often became abusive and unable to provide for their families, women organized and joined the WCTU to oppose and end such abuse. Hundreds of thousands of WCTU women worked for change in a broad spectrum of areas. In addition to suffrage, they sought property rights for married women, the right to custody of children in divorce, and other reforms to assist impoverished and abused women and children; further, they were instrumental in raising the age of sexual consent in nearly every state.
>
> Mattingly 2008, 134

By 1900, WCTU had over 150,000 members nationally and was one the nation's largest women's organizations, attracting largely white Protestant, middle-class women (Daffner 2005). Their efforts went well beyond education about the evils of drink: in Chicago, they had kindergartens, girls' classes in cooking and sewing, recreational and residential facilities, all prior to the founding of Hull-House.

Maps were used as part of the WCTU campaign for temperance, both conceptually as well as for public campaigns (see Figure 4.5).[16] By 1910, the WCTU was advertising in their "Latest Literature and Supplies," in their magazine the *Union Signal*, copies of a "Prohibition Map of the United

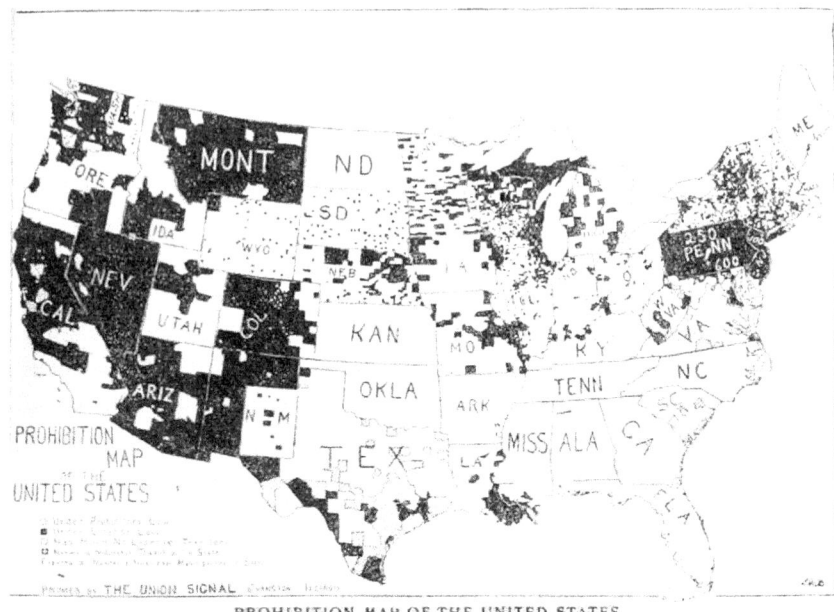

PROHIBITION MAP OF THE UNITED STATES
(Revised January, 1911)

Figure 4.5 Woman's Christian Temperance Union map from the WCTU magazine *Union Signal* 37, 8 (23 February 1911). White is used to indicate areas "Under Prohibition Law" and black those areas "Under License Law." Later maps would simplify it to white as "dry" and black as "wet."

States."[17] The annual Report of the Annual Convention of the National Women's Christian Temperance Union provides state by state accounts of their activities, including the number of maps distributed (Report 1913, 267). A "prohibition map" supposedly held a prominent position on the wall of the National Women's Christian Temperance Union president's office in Evanston IL, charting their progress (Gordon 1924, 263).[18] A song "Make the Map All White" was part of a WCTU songbook:

> O comrade dear, we soon shall hear the good news going round,
> No more in this fair land of ours shall the saloon be found;
> Let us do our part to usher in this day of pure delight
> And together heart and hand to make the map all white ...
>
> Hoffman 1916, 117

A description of a WCTU pageant in Lexington Kentucky describes how they created a "living map" composed of young women in white and black representing states with ("dry") and without temperance ("wet"), who also sang the song "Make the Map All White" ("Living maps" will be discussed in Chapter 5) (119 College Girls 1909).

Catherine Waugh McCulloch, a prominent woman lawyer and activist in the woman's suffrage movement (Illinois Equal Suffrage Association and the National American Woman Suffrage Association), was based out of Chicago and involved with many organizations with ties to Hull-House. Like many politically active Chicago women, McCulloch graduated from Rockford Female Seminary (Boyer 1971). She then attended Union College of Law in Chicago and was finally admitted to practice law in 1898. McCulloch advocated for greater legal protection of women and families. In her 1912 article "Guardianship of Children," she discusses the legal rights of mothers to their children state by state (McCulloch 1912). McCulloch's article was first published in *Chicago Legal News* in 1912 without a map. A map based on her article was published in *The Delineator* in May 1912, accompanying a call to join "The Home League," with the headline "The Home League. It has begun work. It wants you and every friend of yours as a member. This map shows one reason why," with the aim of making " ... the whole of this map white" (Figure 4.6). They state in the text the map was created from data in McCulloch's article (Hard 1912). The map and McCulloch's original article were also included in *The Woman's Citizen Library* Vol. VII Woman Suffrage (Mathews 1913). Through McCulloch's efforts, Illinois passed a bill "granting women equal rights with their husbands in the guardianship of their children. In 1905 the lawmakers adopted another bill, also from her hand, raising the legal age of consent for women from fourteen to sixteen" (Boyer 1971). While McCulloch did not create the map, the map was a visual summary of the text, conveying the state of women's legal rights state by state and a logical accompaniment to the essay.

The visual power of maps led, not only to Hull-House promoting mapping, but also to a number of Hull-House affiliated organizations encouraging others to map. Not long after the publication of a suffrage map in *The Woman's Journal*, a California suffragist wrote in to urge other suffragists to create their own wall-size versions of the map, directing them where to purchase outline maps and display sticks, and to then watercolor or shade them and "a striking wall map evolved, at small expense.... The use of the map is obvious, both on the wall at Suffrage Headquarters, and as a portable object-lesson to display at meetings" (Park 1908). The quote regarding the "ocular proof" of the suffrage map at the start of this section is another example of suffragists encouraging each other to use maps as part of their campaign work.

A 1919 Young Women's Christian Association brochure is a how-to manual for making exhibit maps about labor issues. The brochure's forward states this had originally been created for YWCA secretaries but:

> ... so many requests for the material have been received from all parts of the country, from colleges, women's clubs, chambers of commerce, and from social workers that it has been thought wise to reprint the directions for making the maps and the key to go with the directions in

Figure 4.6 Map created from McCulloch's essay on "Guardianship of Children" (1912) and published by "The Home League" in *The Delineator* 1912. The map has three categories: dark = "Dark Ages"; shaded = "Twilight"; white = "Modern Times." "Modern Times" were states where mothers had legal rights to their children; "Dark Ages" were states where fathers were given sole rights to children regardless of the circumstances. When the map was republished in *The Woman Citizen's Library* (1913), the categories were changed to: "White states have joint guardianship of children. Shaded states give limited power of guardianship to mothers. Black states give practically absolute power of guardianship to fathers."

pamphlet form for sale at a nominal price…. The maps are in a convenient form and are adaptable to many uses.

Directions include suggestions for color schemes, such as "in each map blue indicates the states having the best legal provision for women workers, while white indicates a total absence of legislation, and the sequence of colors indicates relative legal attitudes of states toward women workers"

(YWCA 1919). Much of the brochure provides the information for maps of women's labor with categories and colors provided for all, a sort of color by numbers approach. Outline maps were available for purchase from the brochure's publisher "The Womans Press" of New York, the publishing wing of the YWCA: "Twelve Skeleton Maps Alone" with postage 20 cents.

Anna Nicholes, in her essay "How Women Can Help in the Administration of a City," part of *The Woman Citizen's Library*, calls on women to take action in improving their cities, using techniques the Woman's City Club had employed, especially using community study to determine their needs, then developing exhibits using "posters, maps, and photographs, with short, crisp printed statements showing city problems" to educate the general population (Nicholes 1913, 2154).

There is a very strong spatial component to the work associated with Hull-House and its vast network of associates. They are very much concerned with the conditions in specific locations and how these conditions vary from place to place.

Hull-House and geography

> Young man, I do not believe in geographical salvation.
>   Jane Addams quoted in Berry 1962, 190

While Hull-House residents were studying neighborhoods, mapping them, and discussing them over coffee or dinner and essentially thinking spatially about their community, there were outright ties to the discipline of geography. Geography in the United States in the early 1890s was just becoming established in American universities. Harvard was the first university to offer physical geography in 1871. William Morris Davis was hired by Harvard to teach geology in the 1880s and would go on to become the catalyst for geography's American establishment in the late nineteenth and early twentieth centuries, building Harvard's program, training many physical geographers, and eventually founding the Association of American Geographers in 1904 (Koelsch 2002, 249 and 252). The first geography department in the United States would be established at the University of Chicago, not far from Hull-House, in 1903 (Geography … 1903; Pattison 1981, 157). Jane Addams in particular had many ties to the discipline of geography.

Addams knew Rollin Salisbury, the first chair of geography at the University of Chicago, from her college days in Rockford, with Salisbury attending Beloit College, the brother institution of Rockford Female Seminary and rooming with Addams step-brother George (Brown 2004, 73; Visher 1953, 11). Addams' nephew James Weber Linn states in his biography of Addams that Salisbury proposed to Addams and she refused, a story often repeated in biographies using Linn as a source (Diliberto 1999, 76; Linn

1935, 49).[19] More recently, scholars have suggested the proposal story may be a fabrication by her family to counter speculation about Addams' sexuality (Knight 2010, 184). While the proposal may be questioned, a letter dating from 1881 from Salisbury to Addams in Addams' papers supports they did indeed know each other, but it is a formal, distant letter inviting her and other Seminary students to a lecture at Beloit.

In fact, Salisbury and Addams may have shared an interest in Geography. Salisbury's interest in physical geography/geology was certainly present during his days at Beloit. After his graduation from Beloit, Salisbury was asked to stay on and become a member of the faculty teaching physical geography/geology, using his summers to begin his field work in New Jersey (Densmore 1931, 45–6; Visher 1953, 11). Rockford Female Seminary included geography as part of its curriculum for girls. The curriculum at many Illinois women's colleges and seminaries was based on that of Mount Holyoke; Mount Holyoke's curriculum included ancient geography and physical geography/geology (Enggass 1988). Mcclelland (1943) shows geography was part of the curriculum of several early Illinois women's institutions, and Koelsch (2013) has discussed the teaching of ancient geography, amongst other forms of geography, more widely in nineteenth-century America. In 1878, Jane Addams participated in a state collegiate speech competition and the speech she presented there, "Illinois Geography," discussing the physical geography of Illinois, can be found among her papers (*The Jane Addams Papers* 1985–1986).

Once the work at Hull-House began, their work and Jane Addams' interests led to acquaintances with leading intellectuals around the world, including early social geographers. The Scottish geographer/sociologist Patrick Geddes visited Hull-House several times. A *Chicago Tribune* article covers his visit to Hull-House in April 1899 and his presentation at the Arts and Crafts Society which met at Hull-House (Geddes at Hull-House 1899). Geddes also presented a lecture on "Sex in Education" in February 1900 at Hull-House (Hull-House 1900, 1). Geddes was trained as a botanist but found his passion in the developing field of social sciences, focusing on society and environment (Stevenson 1978, 54). Geddes' interests were extremely broad and extended from studies and theoretical development to exploring alternative educational methods. At the core of Geddes work was the civic or social survey; that in beginning any work, the starting point was a "panoramic view of a definite geographic region" (Geddes 1905, 104–5), followed by a historic survey then a social study, with a strong emphasis on visually presenting the information (Hewitt 2012, 251). Thus, geography was integral to Geddes' approach to sociology. Geddes was friends with and deeply influenced by Pyotr Kropotkin, another friend of Hull-House (Stevenson 1978, 57).

Pyotr Kropotkin, the Russian anarchist geographer, visited Hull-House in 1901, staying a week at Hull-House while giving lectures in the Chicago area on "mutual aid," the subject of his latest book (Avrich 1980, 27; Eddy 2010, 21). Kropotkin shared with Jane Addams an interest in politics and moral development, particularly in the concept of "mutuality." Addams believed humans had an intrinsic "primordial altruism," that "the human urge to provide for and nurture their children and their community in the earliest tribal communities ... predated the beginnings of militarism" (Eddy 2010, 31). In her writings, she linked democracy to this altruism:

> ... she thought that Chicago sat on the cusp of developing a truly democratic industrial community. Once factory workers and factory owners began to see themselves as fellow citizens responsible to each other, they would become a productive human community where the finest fruits of democracy could flourish.
>
> Eddy 2010, 38

Kropotkin believed "mutual aid" or cooperation was key to the higher development of the human species and this cooperation could even be found in the natural world with the most "intelligent species ... ants, parrots, and chimpanzees ... were social animals who aided their own in the struggle for survival" (Eddy 2010, 27). Given Hull-House was known for its salon atmosphere, where discussions over current events, philosophy, and other topics occurred over meals, in the evenings in the parlor, and even after hours in the residents' private spaces, we have to assume Addams and Kropotkin likely had some sort of exchange of ideas at one point or another.

Rollin Salisbury began teaching at the University of Chicago in 1892, teaching "geographic geology." As he built the geography program, Salisbury pulled together the geographical elements being taught in other departments such as botany and zoology, sociology and anthropology, and political economy (Pattison 1981, 155). He hired faculty, some former students of his, who were dedicated teachers, such as J. Paul Goode, Harlan Barrows, Wallace Atwood, and Ellen Churchill Semple (Pattison 1981, 156). Goode was hired in 1903, teaching courses on "Economic Geography of (1) North America, (2) Europe, and (3) Tropical Countries. The central theme of these courses will be the influence of these lands on their settlement, development, and present commercial and industrial status" (Geography ... 1903, 208).

The exact relationship between Hull-House and the geographers of the University of Chicago has yet to be determined, but there *was* a relationship. Through reviewing Hull-House publications and correspondence, we can place geographers within Hull-House and in-relation to Jane Addams. Addams and the residents drew on University of Chicago faculty from many departments to give presentations at their lecture series, including many of the geography faculty. Salisbury presented a stereopticon lecture on "Arctic Exploration" at Hull-House in March 1896 and another on Greenland in February 1897 (Hull-House 1896, 1, 1897, 1).[20] In 1904, Goode gave a lecture series on "Economic Geography" and in 1905 on "Our Natural Resources—Their Economic Significance" (Hull-House 1903–1904, 1, 1904, 1).[21] In December 1906, Goode presented three Sunday lectures (Hull-House 1906–1907, 6). Goode also presented a lecture series on Physical Geography at the Chicago

Commons, the settlement house affiliated with the University of Chicago (Hill 1938, 124).[22] Wallace Atwood, who received his Ph.D. in Geography working with Salisbury at the University of Chicago and became a faculty member in Geology, gave a lecture series on "Scenic Features of North American from a Physiographic Point of View" in 1905.[23] In 1909, he presented an "illustrated" lecture on "The High Mountains of North America" (Hull-House 1910, 7). These lectures were part of an on-going lecture series held at Hull-House, often on Sunday nights, and open to the Hull-House community.

Zonia Baber presents a different interaction with Addams and Hull-House. While she taught at the University of Chicago, her position was in education. I have not located any evidence of her lecturing at Hull-House yet she appears to have had the most interaction with Hull-House and its network of Progressive women, as a Progressive woman herself. As a member of the University of Chicago faculty, Baber undoubtedly knew other female faculty such as Sophinisba Breckenridge and Edith Abbott. In addition to her teaching work, she was also involved with a number of organizations that overlapped with the interests of many Hull-House residents: such as the Chicago Women's Club, the National Association of the Advancement of Colored People, suffrage organizations, and the Women's International League for Peace and Freedom" (Monk 2008). Baber worked with Jane Addams extensively with the Women's International League for Peace and Freedom (ILPF) (Monk and di Friedberg 2011, 76). As part of her work for ILPF, Baber traveled to Austria, Haiti, India, and throughout South America and Africa. Baber's work represents the efforts of Progressive Era American women to extend their work beyond U.S. boundaries to the rest of the world. She shared with Addams a great concern over the future of their world and both worked towards a fairer, more just world for all.

In considering the interaction of Hull-House with early practitioners of geography, we can say the relationship between the Department of Geography at the University of Chicago and Hull-House appears to be cordial but without substantial impact in either direction. So where do we place Hull-House in the development of geography? The most likely place is in social geography – examining "the social contexts, social processes and group relations that shape space, place, nature and landscape" (Ley 2009). Susan Smith et al. in their introduction to the *SAGE Handbook of Social Geographies* credit Élisée Reclus with coining the term "social geography" in approximately 1895, acknowledging the overlap of social geography, social science, and sociology in their development (Smith et al. 2010, 1 and 23). However, in the development of social geography, "radical thinkers" were excluded from the academy by both geographers and sociologists, including Reclus and Kropotkin, as well as the systematic marginalization of the work of Addams/Hull-House and Du Bois by the highly influential Chicago School of Sociology (Smith et al. 2010, 23–25). Sociologists, over the past forty years, have demonstrated the practices of social surveys and mapping associated with Sociology at the University of Chicago have their roots at Hull-House and at other social settlements (Deegan 1981, 1988a; Lengermann and Niebrugge-Brantley 2002; MacLean and Williams 2012; Platt 1994;). Geographers in considering the development of social geography acknowledge the contributions of Addams and Du Bois and their contributions to urban social geography (Del Casino 2009, 38–40; Sibley 1995a, 1995b, 1995c; Smith et al. 2010, 24–5). The practices of social survey with its emphasis on field work and Hull-House's pride in its reports being produced by individuals who know the neighborhood and their neighbors, in many ways has parallels with Carl Sauer's emphasis on field work, despite his "general disdain for 'urban' and modern' places" (Del Casino & Marston 2006, 1005). It appears at least some of the roots of social geography may in fact be in Hull-House.

Matthias Gross' work on early geography and sociology uses the concept of "catchment area" to explore the early development of these disciplines, with catchment area referring to "the domain that members of a discipline declare to be their turf, in which they can legitimately collect data, use methods, or refer to theoretical models that belong to their discipline" (Gross 2004, 576). The overlap in catchment areas between sociology and geography included an emphasis on the relationship of humans to their environment, the study of human societies in general, the practice of field studies, and visualizing human activities using maps. In the 1890s and in the early twentieth century, geography and sociology shared the field of "human ecology." Writing in 1908, sociologist E.C. Hayes described the relationship between sociology and geography as "allies" (Hayes 1908, 39). Interestingly, German geographer Friedrick Ratzel's first published work in English (translated by Ellen Churchill Semple), was in *The American Journal of Sociology*, the premier journal of sociology which often published the social surveys of the settlement houses (in three parts: November 1897, January 1898 and in November 1898). At one point, there was a clear overlap between the academic disciplines of geography and sociology. We also have a third player in this game – social settlements, particularly Hull-House – who also have a claim to this catchment area.

But there is a clear distinction between the players and their approaches to their work. Settlement houses saw their studies as part of a quest for a solution from within the communities themselves. At the beginning of this section, the quote from Addams regarding "geographical salvation" was in response to a running debate about social settlements and their relationships to their communities. "Geographical salvation," according to Addams, is the notion individuals can find a better life simply by moving to the right location, whether it be rural citizens moving to a town for "more fullness of life" or city dwellers moving out of a neighborhood as the neighborhood shifts socio-economically downwards, usually by new immigrants moving in (Addams 1899, 99). Robert Woods, who worked with South End House in Boston, saw social settlements as "laboratories," writing an article on the subject in 1893, where could be found "in essence all the problems of the city, the State, and the Nation" (Wood quoted in Davis 1994, 76). Woods' perspective is not far from that of Ernest Burgess, one of the founders of modern sociology, who wrote in 1916, of the value of social surveys to sociology departments (Burgess 1916). Burgess argues of the value of such work to sociology students, with communities benefitting from "the expert service of the sociologist in making an inventory of its conditions and needs" (Burgess 1916, 494). In contrast, Addams saw communities as having latent potential, with settlements entering the community "... to bring into it and develop from it those lines of thought and action which make for the 'higher life'"(Addams 1899, 55). Addams saw settlements' "... most valuable function as yet, lies along the line of interpretation and synthesis" (Addams 1899, 55). She saw all neighborhoods and their citizens as having the potential to be developed, as not being locked into their destinies.

Besides difference in philosophy of social survey work, even the very nature of the data collection was in question: "Collecting quantitative data was considered 'women's work' by the University of Chicago's male sociologists prior to Ogburn's introduction to the staff in 1928" (Deegan 1988a, 46). In the late nineteenth century, painstaking, tedious work gathering statistics was considered to be feminine work (Mark 1988, 24 and 76). Booth's assistants on *Life and Labour* were primarily women (MacLean and Williams 2012, 246). The practice of visually representing the data in map form was largely associated with Chicago women activists. The quantitative data and the charts and maps created from "women's work" in the settlement houses were viewed as essential to discussing the neighborhood and determining what was needed to address any perceived problems. It was not done for academic ends, but for practical ones.

The pioneering work of the Hull-House and settlement houses in systematically collecting and presenting data on communities was described by Chicago School of Sociology member Ernest Burgess, writing in 1916, as providing "social studies of permanent importance" (Burgess 1916, 493). But their innovative work was not recognized by all, and even Burgess' recognition of the Hull-House work eventually vanished. By the 1920s, the Chicago School would claim social surveys as its own, pointedly ignoring the work of Hull-House residents and its community (Deegan 1988a, 46; Platt 1994, 58). In their claiming of social surveys, there was an academic distancing between the scientists and their objects of study, with the social scientists relying significantly on secondary sources for their data (Platt 1994, 69). By 1926, Park had delineated differences between sociology and geography so that no overlap in catchment area existed in his conceptualization of sociology: there was a clear division of labor between geography and sociology with geography classified as a "nomothetic natural science" and sociology as an "ideographic historical science" (Gross 2004, 593).

Implications of settlement house geography and mapping

The "woman's work" done by Hull-House and other settlement houses in-cluded both "municipal housekeeping" AND the community studies and mapping. In their municipal housekeeping efforts, and in finding meaning-ful work for themselves, they engaged in a practice of social as well as public geography.

This practice of public geography involved the production of knowledge about their community, the immediate community of their neighbor-hoods, the broader urban communities they belonged to, the community of American women, and even extending further to the global level. They also made this information available to their communities – through presenta-tions, through publications, and sometimes just by posting their map on the wall – with the hope they as a community could use this knowledge for their improvement.

The maps they created are essentially thematic maps, each "concerned with a particular subject or theme, such as disease ... many of which are invisible" (Robinson 1982, x). With some noted exceptions, such as the Hull-House maps, the maps themselves are not terribly sophisticated, relying on rather simple dots or hatching to indicate categories. Many of the maps were created by women active in the organizations, such as Agnes Holbrook, Estelle Hunter, and Anna Nicholes, while other maps were drafted by men from data collected by the women, such as *The Pittsburgh Survey* maps, drafted by Shelby Harrison. They employed rather simplistic means to convey their data but understood at least the importance of scale, often em-ploying it quite persuasively, such as the maps depicting conditions within the city block level (such as Figures 4.1 and 4.2). At other times, they used city-wide maps to convey the spatial extent to phenomena such as disease, certain populations or the locations of industries. Finally, they employed national maps to allow viewers to consider the national state of affairs and compare their state to other states, such as Figures 4.4, 4.5, and 4.6. These are not beautiful maps but they serve the purposes of their creators: to com-municate spatial information to their audiences.

Settlement houses were very important locations in knowledge produc-tion. But Addams and other members of the network always stressed their knowledge was situated, it was a view from a specific location (Livingstone 2003, 184). This claim of situated knowledge was used to emphasize their knowledge of the neighborhood or even of their cities was based on first-hand experience, of time spent in the community, and on in-depth surveys of the community. This was to distance themselves from academic sociologists who often viewed their subjects from a distance and who may not have known their subjects as well as they think, relying on secondary data (Platt 1994, 61).

While the community knowledge was situated, it was also presented as rigorous and scientific. At Hull-House, Addams and the residents played this two ways: (1) we are not academics, we know the situation because we live here; YET (2) we use scientific methods to gather statistical data about our community to back up our claims so it has value.

Finally, there is also the knowledge created and circulated by the settlement houses was viewed as being for "greater good" – municipal housekeeping – that is elements of the public realm having an impact on what was considered women's work, such as what's best for children be it education or playgrounds, health issues about water or milk or disease, community welfare and morality. As women whose place in society was limited, this knowledge claim was one men found difficult to argue with. Societal roles of the age constructed a world where men were associated with the public sphere, women with the private. By making claims about the intersection of the public AND the private – women were making a place for themselves.

Reconsidering women's work

The parallels between missionary work and settlement house work were not lost to the women of the Progressive Era. Writing in 1894, Mary Porter commented on social settlements in an article in the *Heathen Woman's Friend*:

> In watching the unfolding and expansion of this work from its germinal thought, "That social interaction could best express the growing sense of the unity of society," I have been constantly impressed by its resemblance in many of its phases to that done on foreign mission fields. It need hardly be said that the prime work of the Christian Missionary is to make known directly and clearly the Gospel of the Lord Jesus Christ.
>
> Porter 1894, 110

Porter goes on to illustrate for her readers how mission work is very much like social settlement work, emphasizing that the social interaction is as important as preaching the gospel:

> Lecture courses on history, science and travel attract many to mission chapels, who do not seek them for religious instruction, and the stereopticon teaches many a delightful lesson, not only in such places, but in remote rural districts where the crowd of spectators gather out of doors and see the pictures thrown upon a screen which covers the outside wall of an adobe house.
>
> Porter 1894, 110–111

By attracting visitors through lectures, settlement houses as well as missions were enticing community members into their spheres and perhaps opening their minds/hearts to their messages.

Both missionaries and settlement workers promoted spatial knowledge and map use. Missionaries, in the course of garnering support for their work, crafted geographies of the missions, both at home and overseas. They wrote and distributed geographies and then advocated for greater geographical awareness and map use. While the majority of the maps they used were basic reference maps they did not create, their use of maps, their endorsement of maps, their insistence on their absolute necessity, and their directions for creating map exhibits in connection with mission studies, promoted cartographic culture among American women. They were producing and circulating information for other women, largely focused on women's issues. In doing so, they were writing from positions of authority and expertise, giving their first-hand experiences in these countries, creating "an image of 'reality' of foreign lands," (Cramer 2003, 212).

Settlement house workers practiced mapping as part of their social survey work, using the maps as persuasive visual evidence of the need for reforms. In many cases, the workers were generating original spatial information, in a practice predating the first department of sociology and geography in the culture. Their "enchantment" with maps *perhaps* bubbled over into other areas of activism, as complementary groups adopted mapping and employed it similarly to visually support their arguments. An important aspect of their use of maps was the association of maps with power and with science. Maps are both a valuable communication tool but also are imbued with power by their viewers, with many readers having blind trust in the information presented on maps (Bar-Gal 2003, 2; Boggs 1947, 469). Maps were both a tool to illuminate community needs AND a persuasive tool to advance their agendas. As settlement houses worked to improve conditions, as organizations such as the Woman's Christian Temperance Union or National Woman's Trade Union League sought to change mindsets and laws, they all wielded maps to serve as "ocular proof" of their claims and to persuade viewers of the validity of the claims.

In gathering support for mission work, in educating their congregations, women's groups, and Sunday school classes about the countries where mission work was occurring, and for social settlements, in studying their communities to determine community needs, they promoted both cartographic culture as well as the practice of cartography. They were not just referring to maps, they were actively engaging with maps in a variety of ways and in a variety of settings. Maps were not just to be glanced at, they were to be actively engaged with following missionaries movements on the map as they read, wall maps were to be not wall paper backdrops but to be pointed at and discussed while studying missionary work, maps were to be crafted to be used with specific missionary lessons. With settlements houses, maps were created to better the neighborhood. These maps might be tacked to a wall and available for all to view and discuss in the settlement houses by residents and neighborhood citizens. They may be used in publications. With organization maps, they may be used in publications or on posters or displays at public forums.

Whether we are considering missionary women or settlement house residents/workers, neither considered themselves geographers, despite their practices of promoting spatial understandings through research, education, and dissemination of information.[24] And in many ways, it does not matter to me whether they did or did not. What is more significant is their contributions to the geographical imaginations of Progressive Era Americans. "Woman's work" led to women needing and using more maps and even creating and using maps in their own work. American women were creating this knowledge largely for other women as part of their efforts to improve their world, especially for other women and girls. They generated this knowledge, they published it, they disseminated it. While not traditionally recognized as part of the geographical canon, it likely had a much great impact, reaching far greater numbers than what was being published by professional geographers of the age.

Notes

1 I live in Omaha, Nebraska: Omaha had a social settlement, the Social Settlement House of Omaha, in South Omaha not far from their Stockyards, established in 1908. It essentially is still in existence today under the name "Kids Can Community Center" (Gebhardt 2011).

2 While university settlements were founded and run by university men, their staff and "workers" were predominantly female. For example, according to a Northwestern University Settlement circular dated 1896, their officers and board members were overwhelming male with eleven men and three women. However, their residents were the reverse: eight women and two men (*Northwestern University Settlement* 1896). At the University Settlement Society of New York, there was a similar hierarchy: the presidents, vice presidents, and council members were all male. Their residents were almost entirely male – seventeen to one woman (a wife). But then they list "workers" which were forty women and fourteen men. They did have a "Women's Auxiliary of the University Settlement Society" but the division of duties is unclear.

3 I will be using "Hull-House" with a hyphen. This was how Jane Addams referred to their venture; thus, it will be my preference.

4 On correspondence between Addams and Gandhi, see: www.mkgandhi.org/articles/adamsgandhi.htm, accessed 15 April 2016.

5 Although, as Kathryn Sklar has noted, Toynbee Hall excluded women (Sklar 1991, 122).

6 The maps were displayed at Toynbee Hall in 1888; Jane Addams visited Toynbee Hall in 1888. We are left to wonder if Addams saw the maps when she was in London, either at Toynbee Hall or elsewhere. In a letter she wrote to her sister on 14 June 1888, Addams discusses briefly her visit to Toynbee Hall. She does not mention Booth's maps in this letter. For the text of the letter, see: www.janeaddamsproject.org/Jane Addams Draft_files/Page 3700.htm, accessed 8 March 2016.

7 In constructing this table, I began with the Jane Addams Papers microfilm and its twenty-three investigations. I then combed through *Hull-House Bulletins* and *Yearbooks*, as well as descriptions of Hull-House written by residents, looking for references to investigations they were conducting. Wherever possible, I located additional citations of the investigations and the published versions of the studies.

8 Samuel Greeley attended the annual "Old Settlers' Party" held at Hull-House on January 1st, 1902 and "gave some extremely interesting reminiscences" according to the 1902 *Hull-House Bulletin*.

9 Repeated attempts to locate this watercolor have been unsuccessful. At the time of exhibition at Hull-House, the water-color was owned by Jane Addams (according to the exhibition catalogue) but I've found no reference to it in the past 100 years.

10 I have not been able to locate a second Hull-House nationalities map to date.

11 I am considering these as "unpublished" as I have searched for these studies and not located published versions (and in fact, only found a manuscript for one of the four studies). They *may* have been published in some form but may be alluding identification.

12 Emphasis is Deegans.

13 The copy of Simkhovitch's *The Settlement Primer* I examined is available through the Hathi Trust Digital Library: http://babel.hathitrust.org/cgi/pt?id=mdp.3 9015005273860;view=2up;seq=6. Interestingly, the copy digitized has a pencil checkmark beside "What to Avoid" in the Table of Contents.

14 In the brief "Map Making" bibliography at the end of the chapter, the references do not include any geographers or cartographers, rather they refer to social science research, statistics, and engineering.

15 Willard was also active in the suffrage movement and quite publicly took up bicycling, publishing *A Wheel Within a Wheel* in 1895.

16 To date, I have not determined whether it was the suffrage movement or the temperance movement who began the trend of using persuasive maps and the phrase "make the map white." Both used maps and the phrase in their campaigns.

17 Listed in every issue from 1910 and 1911 of the *Union Signal*, the WCTU magazine.

18 I have not yet been able to locate a copy of this map but its existence is documented in Gordon 1924, 263. It also comments on the "prohibition map" growing from black to white.

19 It should be noted, whether Linn's story is true or not, Salisbury never married (Visher 1953, 11).

20 James Weber Linn, Addams' nephew, in his biography of her claims after his "proposal" was rejected, Salisbury never entered the doors of Hull-House (Linn 1935, 50). The listing of his lectures in the *Hull-House Bulletin* seems to suggest otherwise, although we have no way of knowing if Salisbury appeared to give the lectures he was scheduled to give. Salisbury also lectured at The University of Chicago Settlement in 1901 (McDowell 1901, 24).

21 The "illustrated lecture" series on Economic geography ran from January 10th to February 14th with lectures on: "The Iron Industry," "The Lumber Industry, "Our Largest Cereal Crop – Corn," "The Economic Significance of the Great Plains," "Cotton as a Social Factor," and "The Reclamation of the Arid Lands" (*Hull-House Bulletin* 6, 1: 1). The following year, Goode's illustrated lectures were on "Our Natural Resources – Their Economic significance" with lectures on "Geographic Location as a Factor in Commercial Supremacy," "The Evolution of a Continent," "The Age of Coal," "When the Coal is Gone, What Then?," "Wheat and Flour," and "The Reclamation of the Arid Lands" (*Hull-House Bulletin* 6, 2: 1). The Hull-House Bulletins generally list several months' worth of Sunday evening lectures but only one or two lecture series. Other lecture series topics include a set of lectures on Industry by a Mrs. A. M. Simons and a set by a Prof. Raymond on the "Capitals of Europe."

On Goode's death in 1932, a letter from Charles Hull Ewing to Jane Addams uses the phrase "our good friend" – "Did you notice that the funeral of our good friend, Dr. J. Paul Goode, was held at the University yesterday?" (Charles Hull Ewing to Jane Addams, 1932 August 10, JAP). Unfortunately, there is no direct correspondence between Goode and Addams in Addams' papers to shed light on this friendship.

22 The reference is part of an essay by Herbert Phillips on Mary McDowell and the work of the University of Chicago Settlement and does not provide a specific date or year when Goode gave his lecture/s.
23 Atwood would go on to teach at Harvard and then Clark University, founding Clark's Graduate School in Geography and becoming Clark's President (James and Martin 1981, 318).
24 Cheryl McEwan in *Geography, Gender and Empire: Victorian Women Travellers in West Africa* observes many of these women travelers, while engaging with and constructing geographic knowledge, did not consider themselves geographers either (McEwan 2000, 5–6 and 216). McEwan goes on to say, "A critical feminist approach need not concern itself with such myopic considerations."

Primary Sources

119 College girls. *Lexington Leader* 5 October 1909.
Abbott, E. 1950. Grace Abbott and Hull-House, 1908–21 Part I. *Social Science Review* 24: 374–394.
Abbott, G. 1909. A study of Greeks in Chicago. *American Journal of Sociology* 15: 373–393.
———. 1915. The midwife in Chicago. *American Journal of Sociology* 20, 5: 684–99.
Adams, M. 1905. Children in American street trades. *Annals of the American Academy of Political and Social Science* 25, 3: 23–44.
———. 1907. *Newsboy Conditions in Chicago*. Chicago, IL: The Federation of Chicago Settlements.
Addams, J. 1899. A function of the social settlement. *The Annals of the American Academy of Political and Social Science* 13: 33–55.
———. 1906. Jane Addams's own story of her work the first five years at Hull-House. *Ladies Home Journal* 23, 5 (April): 11–12.
———. 1930. *The Second Twenty Years at Hull-House: September 1909 to September 1929*. New York: Macmillan.
Addams, J. and A. Hamilton. 1908. The 'piecework' system as a factor in the tuberculosis of wage workers. *International Congress on Tuberculosis Transactions* 3: 139–140.
Addams, J. and H. Grindley. 1898. *A Study of the Milk Supply of Chicago*. Urbana, IL: University of Illinois Agricultural Experiment Station.
Andrews, I. 1917. A black spot off the map. *Life and Labor* 7, 4: 59.
Atwater, W. and A. Bryant. 1898. *Dietary Studies in Chicago in 1895 and 1896*. Washington, DC: Government Printing Office.
Berry, M. 1962. The contribution of the neighborhood approach in solving today's problems. *Social Service Review* 36, 2: 189–193.
Bjorkman, F. and A. Porritt, eds. 1917. *Blue Book. Woman Suffrage: History, Arguments, and Results*. New York: National Woman Suffrage Publishing Co.
Books and Authors. 1895. *The Outlook* 51, 19: 785–786.
Breckenridge, S. and Abbott, E. 1910. Chicago's housing problem: families in furnished rooms. *American Journal of Sociology* 16, 289–308.
———. 1911a. Housing conditions in Chicago, III: back of the yards. *American Journal of Sociology* 16, 4: 433–68.
———. 1911b. Chicago housing conditions, IV: the west side reinvented *American Journal of Sociology* 17, 1–36.

————. 1911c. Chicago housing conditions, V: South Chicago at the gates of the steel mills. *American Journal of Sociology* 17, 145–176.

————. 1912. *The Delinquent Child and the Home*. New York: Charities Publication Committee.

————. 1936. *The Tenements of Chicago, 1908–1936*. Chicago, IL: University of Chicago Press.

Britton, G. 1906. *An Intensive Study of the Causes of Truancy in Eight Chicago Public Schools Including a Home Investigation of Eight Hundred Truant Children*. Chicago, IL: Hollister.

Burgess, E. 1916. The social survey: a field for constructive service by departments of sociology. *American Journal of Sociology* 21, 492–500.

Butler, E. 1909. *Women and the Trades Pittsburgh, 1907–1908*. The Pittsburgh Survey, vol. 3. New York: Charities Publication Committee.

Chicago School of Civics and Philanthropy. 1911. *Chicago School of Civics and Philanthropy Bulletin* 1, 7.

The Committee on Midwives. 1908. The midwives of Chicago. *JAMA: The Journal of the American Medical Association* 50, 17: 1346–1350.

Comstock, A. 1912. Chicago housing conditions, VI: the problem of the Negro. *American Journal of Sociology* 18: 241–57.

Densmore, H. 1931. Rollin D. Salisbury, M.A. LL.D., A biographical sketch. *The Wisconsin Magazine of History* 15, 1: 22–46.

Du Bois, W. 1899. *The Philadelphia Negro: A Social Study*. Philadelphia, PA: Published for the University.

Elson, A. 1954. First principles of Jane Addams. *Social Service Review* 28 (March), 3–11.

Geddes, P. 1905. Civics: as Applied Sociology. *Sociological Papers* 1: 105.

Geddes at Hull-House. *Chicago Tribune* 1 April 1899, p. 8.

Geography in the University of Chicago. 1903. *Bulletin of the American Geographical Society* 35, 2: 207–208.

Gernon, M., Howe, G. and Hamilton, A. 1903a. *An Inquiry into the Causes of the Recent Epidemic of Typhoid Fever in Chicago*. Chicago, IL: City Homes Association.

————. 1903b. An inquiry into the part played by the housefly in the recent epidemic of typhoid fever. *The Commons* 8: 83.

Gordon, E.P. 1924. *Woman Torch-Bearers: The Story of the Women's Christian Temperance Union*. Evanston, IL: National Woman's Christian Temperance Union Publishing House.

Hamilton, A. 1903. The fly as a carrier of typhoid. *Journal of the American Medical Association* 40, 9: 576–583.

————. 1905. *A Study of Tuberculosis in Chicago*. Chicago, IL: City Homes Association.

————. 1910. Excessive child-bearing as a factor in infant mortality. *Bulletin of the American Academy of Medicine* 11: 181–187.

Hard, W. 1912. The home league. *The Delineator* 79, 5: 392.

Hayes, E. C. 1908. Sociology and psychology; sociology and geography. *American Journal of Sociology* 14, 3: 371–407.

Henderson, C. 1899. *Social Settlements*. New York: Lentilhon & Company.

Hill, C., ed. 1938. *Mary McDowell and Municipal Housekeeping: A Symposium*. Chicago, IL: Millar Publishing Co.

Hoffman, E. 1916. *Songs of the New Crusade: A Collection of Stirring Twentieth Century Temperance Songs*. Chicago, IL: Hope Publishing Company.

Holbrook, A. 1894. Hull-House. *Wellesley Magazine* 2: 171–80.

Hooker, G. [1896]. *A List of Pleasant Places for Nineteenth Warders to Go ...* Chicago, IL: Hull-House.

Howe, M. 1896. Settlers in the city wilderness. *The Atlantic Monthly* 77, 459: 118–123

Hull-House. 1895. *Hull-House Art Exhibition*. Chicago, IL: Hull-House Association Records.

———. 1896. *Hull-House Bulletin* (March).

———. 1897. *Hull-House Bulletin* (February).

———. 1898. *Hull-House Bulletin* 3: 7.

———. 1900. *Hull-House Bulletin* (Midwinter).

———. 1902. *Hull-House Bulletin* 5, 1.

———. 1903. *An Inquiry into the Causes of the Recent Epidemic of Typhoid Fever in Chicago*. Chicago, IL: City Homes Association of Chicago.

———. 1903–1904. *Hull-House Bulletin* 6: 1.

———. 1904. *Hull-House Bulletin* 6: 2.

———. 1905–1906. *Hull-House Bulletin* 7: 1.

———. 1906–1907. *Hull-House Year Book*. Chicago, IL: Hull-House.

———. 1910. *Hull-House Year Book*. Chicago, IL: Hull-House.

———. 1929. *Hull-House Year Book*. Chicago, IL: Hull-House.

———. 1934. *Hull-House Year Book: Forty-Fifth Year*. Chicago IL: Fred Klein.

Hull-House Maps and Papers. *The Outlook* 51, 19: 785–6.

Hull-House Residents. 1903. An inquiry into the causes of the recent epidemic of typhoid fever in Chicago. *The Commons* 8: 3–7

Hunt, C. 1897. *The Italians in Chicago: A Social and Economic Study. Ninth Special Report of the Commission of Labor*. Washington, DC: Government Printing Office.

Hunter, R. 1901. *Tenement Conditions in Chicago*. Chicago, IL: City Homes Association.

League for the Protection of Immigrants. 1910. *Annual Report*. Chicago, IL: League for the Protection of Immigrants.

Jones, A. 1928. Mexican colonies in Chicago. *Social Service Review* 2, 4: 579–597.

Kawin, E. 1933. Intelligence and poverty. *Survey Graphic* 22, 10 (October): 502–4.

Kelley, F. 1894. *First Annual Report of the Factory Inspectors of Illinois for the Year Ending December 15, 1893*. Springfield, IL: State Printers.

Kellogg, P., ed. 1909-14. *The Pittsburgh Survey; findings in six volumes*. New York: Charities Publication Committee.

———. 1898. Hull-House. *The New England Magazine* 18, 5: 550–566.

Kelley, N. 1954. Early days at Hull-House. *The Social Service Review* 28, 424–429.

Koren, J. 1899. *Economic Aspects of the Liquor Problem*. New York: Houghton, Mifflin and Company.

Linn, J. 1935. *Jane Addams – A Biography*. New York: Appleton-Century Co.

Lloyd, H. [1903?] *The Chicago Traction Question*. Chicago, IL: George Waite Pickett.

Maude, A. 1902. A talk with Miss Jane Addams and Leo Tolstoy. *The Humane Review* 3 (Oct. 1902): 203–218.

McCulloch, C. 1908. Letter to the editor. *The Woman's Journal*, 18 January: 12.

———. 1912. Guardianship of children. *Chicago Legal News* 44, 23 (Jan. 13): 180 and 182.

McDowell, M. 1901. *The University of Chicago Settlement.* Chicago, IL: [s.n.].

Milner, R. 1903. *Dietary Studies in Boston and Springfield Mass., Philadelphia Pa. and Chicago Il.* By Lydia Southard, Ellen H. Richards, Susannah Usher, Bertha M. Terrill, and Amelia Shapleigh. US Department of Agriculture, Office of Experiment Stations—Bulletin No. 129. Washington, DC: Government Printing Office.

Montgomery, C. 1905. *Bibliography of College, Social, University and Church Settlements.* Chicago, IL: Blakely Press.

Montgomery, L. 1913. *The American Girl in the Stockyards District.* Chicago, IL: University of Chicago Press.

Moore, D. 1897a. A day at Hull-House. *American Journal of Sociology* 2: 629–42.

Moore, E. 1897b. The social value of the saloon. *The American Journal of Sociology* 3, 1: 1–12.

Nestor, A. 1910? *The Working Girl's Need of Suffrage.*

Nicholes, A. 1913. How women can help in the administration of a city. In *The Woman Citizen's Library: A Systematic Course of Reading in Preparation for the Larger Citizenship*, vol. IX, 'Woman and the Larger Citizenship: City Housekeeping,' ed. S. Matthews, pp. 2143–2208. Chicago, IL: The Civics Society.

Northwestern University Settlement, 252 West Chicago Avenue, Chicago, IL. Circular No. 6. June 1896.

Palmer, V. 1928. *Field Studies in Sociology: A Student's Manual.* Chicago, IL: University of Chicago Press.

Park, A. 1908. Suffrage map enlarged. *The Woman's Journal* 39, 29: 113.

Porter, M. 1894. The social settlement element in foreign missionary work. *Heathen Woman's Friend* 26, 4: 110–111.

Proceedings of the Thirty-Ninth Annual Convention of the National American Woman Suffrage Association, held at Chicago, IL, February 14 to 19, 1907. Warren, OH: Press of Wm. Ritezel & Co.

Profit in child victims to cocaine. 1904. *The Commons* 9: 423.

Report of the … annual convention of the National Woman's Christian Temperance Union. [1913]. Chicago, IL: Woman's Temperance Publishing Association.

Residents of Hull-House. 1895. *Hull-House Maps and Papers, a presentation of nationalities and wages in a congested district of Chicago, together with comments and essays on problems growing out of the social conditions.* New York and Boston: T.Y. Crowell.

Scudder, V. 1890. The relation of college women to social need. *Association of Collegiate Alumnae Series II* (October 1890): 1–16

Shapleigh, A. 1892–1893. *A Study of Dietaries: Partial Report of Dutton Fellows*, College Settlements Association.

———. 1894. A study of dietaries. *The New England Kitchen: A Monthly Journal of Domestic Science* 1, 4: 203–211.

Simkhovitch, M. 1926. *The Settlement Primer.* Boston, MA: National Federation of Settlements.

Turner, F. 1914. The west and American ideals. *The Washington Historical Quarterly* 5, 4: 243–257.

Van Der Vaart, H. 1902. Child labor in Illinois. *The Commons* 7: 70.

———. 1903. Our working children in Illinois. *The Commons* 7: 79.

———. 1904. Child workers at the holiday season. *The Commons* 9, 2: 57–9.

Walker, N. 1915. Housing conditions, X: Greeks and Italians in the neighborhood of Hull-House. *American Journal of Sociology* 21: 285–316.

"Walks and Wins with Two-Ft. Map." 1913. *The Woman's Journal*, 1 February: 40.

Webb, B. 1963. *American Diary, 1898*. Ed. D. Shannon. Madison, WI: University of Wisconsin Press.

Wescott, O. 1903. The men of the lodging houses. *The Commons* 86, 8: 1–5.

Woods, R. and A. Kennedy, eds. 1911. *Handbook of Settlements*. New York: Charities Publication Committee, The Russell Sage Foundation.

Wright, C. 1894. *The Slums of Baltimore, Chicago, New York and Philadelphia*. Seventh Special Report of the Commissioner of Labor. Washington, DC: General Printing Office.

Young Women's Christian Association of the U.S.A. Industrial Committee. 1919. *State Laws Affecting Women in the United States and Directions for Making Exhibit Maps*. New York: Womans Press.

Bibliography

Avrich, P. 1980. Kropotkin in America. *International Review of Social History* 25, 1: 1–34.

Bales, K. 1991. Charles Booth's survey of life and labour of the people in London 1889–1903. In *The Social Survey in Historical Perspective 1880–1940*, eds. M. Bulmer, K. Bales, and K. Sklar, pp. 66–110. Cambridge, UK: Cambridge University Press.

Barbuto, D. 1999. *American Settlement Houses and Progressive Social Reform: An Encylopedia of the American Settlement Movement*. Phoenix, AZ: Oryx Press.

Bar-Gal, Y. 2003. The Blue Box and JNF propaganda maps, 1930–1947. *Israel Studies* 8, 1: 1–19.

Boggs, S.W. 1947. Cartohypnosis. *Scientific Monthly* 65/1: 469–76.

Boyer, P. 1971. Mcculloch, Catharine Gouger Waugh (June 4, 1862-April 20, 1945). In *Notable American Women: 1607–1950*, eds. E. James, J. James and P. Boyer. Cambridge, MA: Harvard University Press,http://leo.lib.unomaha.edu/login?url=http://literati.credoreference.com/content/entry/hupnawi/mcculloch_catharine_gouger_waugh_june_4_1862_apr_20_1945/0. Accessed 21 April 2016.

Brown, V. 1995. Jane Addams, progressivism, and woman suffrage: an introduction to 'Why Women Should Vote.' In *One Woman, One Vote*, ed. M. Wheeler, pp. 179–202. Troutdale, OR: NewSage Press.

———. 2004. *The Education of Jane Addams*. Philadelphia: University of Pennsylvania Press.

Bryan, M., N. Slote, and de Angury, M. 1996. *The Jane Addams Papers: A Comprehensive Guide*. Bloomington, IN: Indiana University Press.

Bulmer, M. 1996. The social survey movement and early twentieth-century sociological methodology. In *Pittsburgh Surveyed*, eds. M. Greenwald and M. Anderson, pp. 15–34. Pittsburgh, PA: University of Pittsburgh Press.

———. 1991a. The decline of the social survey movement and the rise of American empirical sociology. In *The Social Survey in Historical Perspective 1880–1940*, eds. M. Bulmer, K. Bales, and K. Sklar, pp. 291–315. Cambridge, UK: Cambridge University Press.

———. 1991b. W.E.B. Du Bois as a social investigator: *The Philadelphia Negro*, 1899. In *The Social Survey in Historical Perspective 1880–1940*, eds. M. Bulmer, K. Bales and K. Sklar, pp. 170–188. Cambridge, UK: Cambridge University Press.

Bulmer, M., K. Bales, and K. Sklar. 1991. The social survey in historical perspective. In *The Social Survey in Historical Perspective 1880–1940*, eds. M. Bulmer, K. Bales and K. Sklar, pp. 1–48. Cambridge, UK: Cambridge University Press.

Buroker, R. 1971. From voluntary association to welfare state: the Illinois Immigrants' Protective League, 1908–1926. *Journal of American History* 58: 643–660.

Cohen, S. 1991. The Pittsburgh survey and the social survey movement: a sociological road not taken. In *The Social Survey in Historical Perspective 1880–1940*, eds. M. Bulmer, K. Bales and K. Sklar, pp. 244–268. Cambridge, UK: Cambridge University Press.

Coohey, C. 1999. Notes on the origins of social work education. *Social Services Review* 73, 418–422.

Cramer, J. 2003. White womanhood and religion: colonial discourse in the U.S. women's missionary press, 1869–1904. *The Howard Journal of Communications* 14: 209–224.

Daffner, N. 2005. *Woman's Christian Temperance Union*. The Encyclopedia of Chicago, www.encyclopedia.chicagohistory.org/. Accessed 29 December 2016.

Davis, A. 1994. *Spearheads for Reform: The Social Settlements and the Progressive Movement 1890–1914*. New Brunswick, NJ: Rutgers University Press.

Deegan, M. 1981. Early women sociologists and the American Sociological Society: the patterns of exclusion and participation. *The American Sociologist* 16: 14–24.

———. 1988a. *Jane Addams and the Men of the Chicago School*. New Brunswick, NJ: Transaction Books.

———. 1988b. W.E.B. Du Bois and the women of Hull-House, 1895–1899. *The American Sociologist* 19, 4: 301–311.

Del Casino Jr., V. 2009. *Social Geography: A Critical Introduction*. Malden, MA: Wiley-Blackwell.

Del Casino Jr., V. & S. Marston. 2006. Social geography in the United States: everywhere and nowhere. *Social & Cultural Geography* 7, 6: 995–1009.

Diliberto, G. 1999. *A Useful Woman: The Early Life of Jane Addams*. New York: Lisa Drew/Scribner.

Dillon, D. [undated] *Agnes Sinclair Holbrook*. http://hullhouse.uic.edu/hull/urbanexp/main.cgi?file=new/show_doc.ptt&doc=280&chap=39. Accessed 29 December 2016.

Diner, S. 1980. *A City and Its Universities: Public Policy in Chicago, 1892–1919*. Chapel Hill, NC: University of North Carolina Press.

Eddy, B. 2010. Struggle or mutual aid: Jane Addams, Petr Kropotkin, and the Progessive encounter with Social Darwinism. *The Pluralist* 5, 1: 21–43.

Enggass, P. 1988. Geography at Mount Holyoke Female Seminary and College, 1837–1984. In *Geography in New England*, eds. J. Harmon and T. Rickard, pp. 25–31. New Britain, CT: New England/St. Lawrence Valley Geographical Society.

Fish, V. 1985. Hull-House: pioneer in urban research during its creative years. *History of Sociology* 6: 33–34.

———. 2001. Sophonisba Preston Breckenridge. In *Women Building Chicago 1790–1990: A Biographical Dictionary*, eds. R. Schultz and A. Hast, pp. 114–16. Bloomington and Indianapolis, IN: Indiana University Press.

Flanagan, M. 2005. Suffrage. In *The Electronic Encyclopedia of Chicago*, eds. J. Reiff, A. Keating and J. Grossman. Chicago, IL: Chicago Historical Society. www.encyclopedia.chicagohistory.org/pages/1217.html. Accessed 5 April 2015.

Gebhardt, L. 2011. *Social settlement and the Omaha community*. M.S. thesis, University of Nebraska Omaha.

Glowacki, P. and J. Hendry. 2004. *Images of America: Hull-House*. Charleston, SC: Arcadia Publishing.

Grant, L., M. Stalp, and K. Ward. 2002. Women's sociological research and writing in the AJS in the pre-World War II era. *The American Sociologist* 33: 69–91.

Greenwald, M. 1996. Visualizing Pittsburgh in the 1900s: art and photography in the service of social reform. In *Pittsburgh Surveyed*, eds. M. Greenwald and M. Anderson, pp. 124–152. Pittsburgh, PA: University of Pittsburgh Press.

Gross, M. 2004. Human geography and ecological sociology: the unfolding of a human ecology, 1890 to 1930—and beyond. *Social Science History* 28, 4: 575–605.

Hewitt, L. 2012. The civic survey of greater London: social mapping, planners and urban space in the early twentieth century. *Journal of Historical Geography* 38: 247–262.

Higbie, T. 2005. Women's Trade Union League. *The Electronic Encyclopedia of Chicago*, http://www.encyclopedia.chicagohistory.org/pages/1373.html. Accessed 19 June 2017.

Hillier, A. 2009. W.E.B. Du Bois and the social survey movement. In *The Shape of Philadelphia*, ed. Cartographic Modeling Lab. Philadelphia, PA: Cartographic Modeling Lab. http://works.bepress.com/amy_hillier/27/. Accessed 27 February 2016.

Horowitz, H. 1983/1984. Hull-House as women's space. *Chicago History* 12, 40–55.

Jackson, S. 2000. *Lines of Activity: Performance, Historiography, Hull-House Domesticity*. Ann Arbor, MI: University of Michigan Press.

Jackson, B. 2005. National Women's Trade Union League. In *Encyclopedia of the Gilded Age and Progressive Era*, eds. J. Buenker and J. Buenker, pp. 715–716. Armonk, NY: Sharpe Reference.

James, P. and G. Martin. 1981. *All Possible Worlds: A History of Geographical Ideas*. New York: John Wiley & Sons.

Katzman, D. 1981. *Seven Days a Week: Women and Domestic Service in Industrializing America*. Urbana and Chicago, IL: University of Illinois Press.

Kimball, M. 2006. London through rose-colored graphics: visual rhetoric and information graphic design in Charles Booth's maps of London poverty. *Journal of Technical Writing and Communication* 36, 4: 353–381.

Kirkby, D. 1992. Class, gender and the perils of philanthropy. *Journal of Women's History* 4, 2: 36–51.

Kleinberg, S. 1996. Seeking the meaning of life: The Pittsburgh Survey and the family. In *Pittsburgh Surveyed*, eds. M. Greenwald and M. Anderson, pp. 88–105. Pittsburgh, PA: University of Pittsburgh Press.

Knight, L. 2005. *Citizen: Jane Addams and the Struggle for Democracy*. Chicago, IL: University of Chicago Press.

———. 2010. Love on Halsted Street: a contemplation on Jane Addams. In *Feminist Interpretations of Jane Addams*, ed. M. Hamington, pp. 181–200. University Park, PA: The Pennsylvania State University Press.

Koelsch, W. 2002. Academic geography, American style: an institutional perspective. In *Geography: Discipline, Profession and Subject since 1870: An International Survey*, ed. G. S. Dunbar, pp. 281–316. Dordrecht: Kluwer.

———. 2013. *Geography and the Classical World: Unearthing Historical Geography's Forgotten Past*. London: I. B. Tauris.

Lengermann, P. and Niebrugge-Brantley, J. 2002. Back to the future: settlement sociology, 1885–1930. *The American Sociologist* 3, 5–20.

Ley, D. 2009. Social geography. In *The Dictionary of Human Geography*, eds. D. Gregory, R. Johnston, G. Pratt, M. Watts and S. Whatmore. Oxford: Blackwell Publishers.

Livingstone, D. 2003. *Putting Science in Its Place: Geographies of Scientific Knowledge*. Chicago, IL: The University of Chicago Press.

MacLean, V. and J. Williams. 2012. "Ghosts of sociologies past": settlement sociology in the Progressive Era at the Chicago School of Civics and Philanthropy. *The American Sociologist* 43: 235–63.

Mark, J. 1988. *A Stranger in Her Native Land: Alice Fletcher and the American Indian*. Lincoln, NE: University of Nebraska Press.

Mathews, S., ed. 1913-14. *The Woman Citizen's Library: a systematic course of reading in preparation for the larger citizenship*. Chicago IL: The Civics Society.

Mattingly, C. 2008. Woman's temple, women's fountains: the erasure of public memory. *American Studies* 49, 3/4: 133–156.

McClelland, C. 1943. The education of females in early Illinois. *Journal of the Illinois State Historical Society* 36: 378–407.

McCree, M., N. Slote, and M. De Angury, eds. 1996. *The Jane Addams Papers: A Comprehensive Guide*. Bloomington, IN: Indiana University Press.

McEwan, C. 2000. *Gender, Geography and Empire: Victorian Women Travellers in West Africa*. Aldershot UK: Ashgate.

McGrail, S. 2013. The Philadelphia Negro. In *Encyclopedia of Greater Philadelphia*. http://philadelphiaencyclopedia.org/archive/philadelphia-negro-the/. Accessed 27 February 2016.

Monk, J. 2008. *Practically All the Geographers Were Women: About the Society of Woman Geographers*. http://iswg.coastalgraphics.com/news-events/practically-all-the-geographers-were-women/. Accessed 23 February 2016.

Monk, J. and M. di Friedberg. 2011. Mary Arizona (Zonia) Baber. *Geographers Biobibliographical Studies* 30: 68–79.

Noblitt, J. with A. Zárasová. 2001. Josefa Veronika Humpal-Zeman. In *Women Building Chicago 1790–1990: A Biographical Dictionary*, eds. R. Schultz and A. Hast, pp. 417–419. Bloomington and Indianapolis, IN: Indiana University Press.

O'Connor, S. 2009. Methodological triangulation and the social studies of Charles Booth, Jane Addams, and W.E.B. Du Bois. *Sociation Today* 7, 1. www.ncsociology.org/sociationtoday/dubois/three.htm. Accessed 27 February 2016.

Ogawa, M. 2004. 'Hull-House' in downtown Tokyo: the transplantation of a settlement house from the United States into Japan and the North American missionary women, 1919–1945. *Journal of World History* 15, 3: 359–387.

Pattison, W. 1981. Rollin Salisbury and the establishment of geography at the University of Chicago. In *The Origins of Academic Geography in the United States*, ed. B. Blout, pp. 151–163. Hamden, CT: Archon Books.

Platt, J. 1994. The Chicago School and firsthand data. *History of the Human Sciences* 7, 1: 57–80.

Porter, J. 2011. Context, location, and space: the continued development of our 'geo-sociological' imaginations. *The American Sociologist* 42, 4: 288–302.

Robinson, A. 1982. *Early Thematic Mapping in the History of Cartography*. Chicago, IL: University of Chicago Press.

Schultz, A. 2013. The black mammy and the Irish Bridget: domestic service and the representation of race, 1830–1930. *Éire-Ireland* 48, 3 & 4: 176–212.

Schultz, R. 2007. Introduction. In *Hull-House Maps and Papers*, ed. R. Schultz, pp. 1–42. Urbana and Chicago, IL: University of Chicago Press.

Schultz, R. and A. Hast. 2001. Introduction. In *Women Building Chicago 1790–1990: A Biographical Dictionary*, eds. R. Schultz and A. Hast, pp. xix–lv. Bloomington and Indianapolis, IN: Indiana University Press.

Sibley, D. 1995a. Women's research on Chicago in the early 20th century. *Women & Environments* 14: 6–8.

———. 1995b. *Geographies of Exclusion: Society and Difference in the West.* London: Routledge.

———. 1995c. Gender, science, politics and geographies of the city. *Gender, Place and Culture* 2: 37–49.

Sklar, K. 1985. Hull-House in the 1890s: a community of women reformers. *Signs: Journal of Women in Culture and Society* 10, 658–677.

———. 1991. *Hull-House Maps and Papers*: social science as women's work in the 1890s. In *The Social Survey in Historical Perspective 1880–1940*, eds. M. Bulmer, K. Bales and K. Sklar, pp. 111–147. Cambridge, UK: Cambridge University Press.

Smith, S., R. Pain, S. Marston, and J. Jones III. 2010. Introduction: situating social geographies. In *The SAGE Handbook of Social Geographies*, eds. S. Smith, R. Pain, S. Marston and J. Jones III, pp. 1–39. Thousand Oaks, CA: SAGE Publications.

Stebner, E. 2006. The settlement house movement. In the *Encyclopedia of Women and Religion in North America*, eds. R. Keller and R. Ruether, pp. 1059–1069. Bloomington, IN: Indiana University Press.

Stevenson, I. 1978. Patrick Geddes 1854–1932. *Geographers Biobibliographical Studies* 2: 53–65.

The Jane Addams Papers. 1985–6. Edited by M. McCree Bryan et al. Microfilm, 82 reels. Ann Arbor, MI: University Microfilms International.

Topalov, C. 1993. The city as terra incognita: Charles Booth's poverty survey and the people of London, 1886–1891. *Planning Perspectives* 8: 395–425.

van den Hoonaard, W. 2013. *Map Worlds: A History of Women in Cartography.* Waterloo, ON: Wilfrid Laurier University Press.

Vapnek, L. 2009. *Breadwinners: Working Women & Economic Independence 1865–1920.* Urbana and Chicago, IL: University of Illinois Press.

Visher, S. 1953. Rollin D. Salisbury and geography. *Annals of the Association of American Geographers* 63: 4–11.

Wade, L. 2001. Mary Eliza McDowell. In *Women Building Chicago 1790–1990: A Biographical Dictionary*, eds. R. Schultz and A. Hast, pp. 563–565. Bloomington and Indianapolis, IN: Indiana University Press.

Williams, J. and V. MacLean. 2012. In search of the kingdom: the social gospel, settlement sociology, and the science of reform in America's progressive era. *Journal of the History of the Behavioral Sciences* 48, 4: 339–362.

———. 2015. *Settlement Sociology in the Progressive Years: Faith, Science and Reform.* Boston, MA: Brill.

5 Changing the map – political activism, geography, and cartography

> ... if woman would fulfil her traditional responsibility to her own children; if she would educate and protect from danger factory children who must find their recreation on the street; if she would bring the cultural forces to bear upon our materialistic civilization; and if she would do it all with the dignity and directedness fitting one who carries on her immemorial duties, then she must bring herself to the use of the use of the ballot—that latest implement for self government. May we not fairly say that American women need this implement in order to preserve the home?
>
> Jane Addams, "Why Women Should Vote" 1910

The practice of "women's work" in the Progressive Era covered a broad range of social issues, many concerning the conditions women endured at home and abroad. As they worked to improve these conditions, the call for women to have the vote was intricately involved in their efforts, with the mindset being that if women were to improve their lot and those of other women and children, they needed to have the vote (Wheeler 1995, 15). If they were going to enact positive change in their communities, if they were to "preserve the home" as Jane Addams suggests (in the above quote), they were going to need more than pens and voices.

This movement to get the vote extended to women was one of many efforts to impact the United States on a legislative level. In their work to improve working conditions or to advocate for women's rights to their children, they too were advocating for legal rights and changes to existing laws and regulations. But the suffrage issue became a massive lobbying effort as women had to convince male voters of the value of women having the vote in order for the U.S. Constitution to be changed (eventually they would see 36 states ratify the Amendment and it passed in both the House and the Senate). In this effort, women made many arguments why they should have the vote – in terms of equal rights (the only men denied the vote were "idiots, lunatics, illiterates, and criminals"), home protection (consumer protection laws, sanitary conditions in cities), and social justice (poverty, child labor laws, temperance, women's legal rights to their children) – but they also made a visual argument, a "suffrage map," that was used eventually national-wide

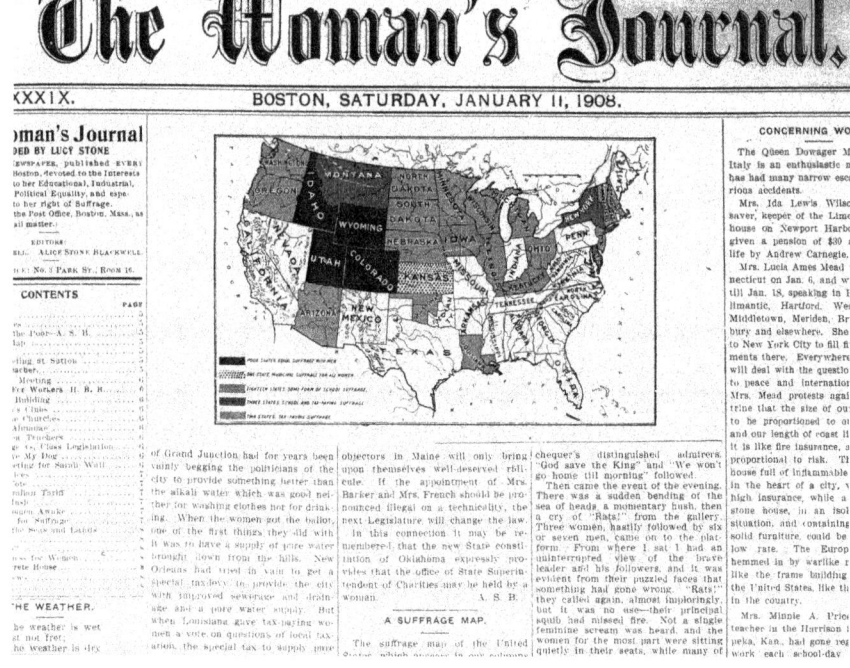

Figure 5.1   Bertha Knobe's map reprinted on the front page of *The Woman's Journal* 39, 2 (11 January 1908): 5.

as they ramped up efforts to obtain suffrage (Figure 5.1) (Knobe 1907, 772). The suffrage map, used from approximately 1908 until passage of the 19th Amendment in 1920, represents the most extensive use of an iconic map image for persuasive purposes in the United States. In this chapter, I discuss the creation and implementation of the suffrage map and how it came to be an icon of the suffrage campaign. This use of a map goes beyond being "ocular proof" or being persuasive to being propaganda. The use of the suffrage map and similar maps, such as those used by the Association of Southern Women for the Prevention of Lynching, were for a different form of women's work: legislative change. Through maps such as these, American women were working to change the American landscape and to *change* these very maps.

   The maps discussed in the previous chapter are largely thematic maps, that is, they "focus on the distribution of social or physical phenomena at relatively small scales" (Slocum and Kessler 2015, 1500). Using relatively simple techniques, such as shading or dots to indicate distributions on simple outline maps (sometimes provided by map companies like Rand McNally), women created maps and used them in their social activism, mapping the social phenomena they were concerned with and then using these maps to persuade their audiences of the need for certain changes, be it better living conditions on Chicago's South Side, an eight-hour work day, or women's legal rights to their children. Writing in 1942, Hans Speier stated:

> Maps enable us to perceive visually relationships within a context, as on a symbolic plane we can see what would not be clear to us if we had to rely on immediate experience and memory. Thus maps help us not only to visualize the meaning of more remote facts, but they also organize details into a whole which, without the help of the map, would be lost because of our own limitations.
>
> Speier 1942, 311

Yet, cartographers have known for a long time of "cartohypnosis" or "hypnotism by cartography": the acceptance "subconsciously and uncritically the ideas that are suggested to them by maps" (Boggs 1947, 469).

With more politically charged issues such as the suffrage map, eventually anti-lynching maps and even in the case of the Mormon octopus maps discussed in Chapter 3, thematic maps depicting the spatial distribution of spatial information were crafted into even more persuasive maps and into the realm of propaganda. I agree with Judith Tyner's assertion that all maps are persuasive, "... used with the intention of influencing the opinions or beliefs of readers" and that there is a continuum in mapping from expository to persuasive cartography (Tyner 2015, 1087). But at least in the case of the suffrage map, I would argue the term propaganda is accurate, if for no other reason than the suffragists included the map in their "propaganda." Propaganda has come to be associated with negatives: "equated with evil—lies, deliberate distortions, misinformation—whereas *persuasion* implied an appeal based on logic or inducement rather than false assertion" (Tyner 2015, 1088).[1] The persuasive mapping done in Germany and Italy around World War II is often seen as the epitome of propaganda mapping (Boria 2008; Herb 1997; Minor 1999). But at the time of the suffrage maps' conception and implementation (approximately 1907–1919), propaganda was not equated with evil and lies. At annual suffrage meetings, such as at the Forty-Fourth Annual Convention of the National American Woman Suffrage Association in 1912, open discussions were scheduled to be conducted on "propaganda" (as well as other topics such as Press Work, Congressional Work and Traveling Exhibits) (National American Woman Suffrage Association 1912, 7). The reports of various committees and of the state committee efforts included in the proceedings of the annual suffrage meetings report on the quantity of propaganda distributed (such as NAWSA 1907, 27–228; NAWSA 1912, 73; NAWSA 1914, 154; NAWSA 1919, 160–1). Hull-House, in reporting on their activities as a social settlement, commented on the value of propaganda in their work and its importance in drawing attention to their community's needs (Woods and Kennedy 1911, 59). Some anti-suffrage activists certainly viewed the suffrage map as propaganda, perhaps not as "evil" and "lies," but used it to argue against suffrage, referencing the spatial patterns. As the suffragists viewed the map as propaganda, I stand with the suffragists and also view it as propaganda.

The suffrage and anti-lynching maps however, also could be seen as early examples of critical cartography, of women taking and exerting power through mapping. Critical cartography can be defined as "the emancipatory and subversive effects of mapping practices (including digital mapping with GIS) that are emerging outside of the cartography traditionally controlled by the state and corporate interests" (Pavlovskaya and St. Martin 2007, 590). While used to describe contemporary women's adoption of cartography to advance their social justice work, I believe it is equally applicable to Progressive Era women's use of cartography to advance their political agendas.

I will begin by discussing the suffrage map, its development, its implementation, its racial implications, and its eventual embodiment by suffragists. I will then consider another example of women-created propaganda maps – antilynching maps. In both cases, women were bent on changing their maps. While they were using the maps to make their cases, mark their progress, and argue for their success, it was ultimately about mobilizing Americans to transform their nation, to *change* the map.

"That suffrage map": The American suffrage movement and their maps

The suffrage movement in the United States is generally viewed as beginning with the Seneca Falls Women's Rights Convention of 1848. In the early days of the movement, the end goal was not suffrage alone, rather suffrage was tied to

> social reform movements, such as antislavery, health, education, and labor conditions.... Having been denied the right to participate fully in reform movements, women began speaking out for changes in the social, moral, legal, educational, and economic status for themselves and others: their initial focus centered on women's rights as a whole, not specifically on woman suffrage.
>
> Jenkins 2011, 133

In the late nineteenth century, suffrage tactics were fairly low-key, holding meetings of sympathetic women and "discretely lobbied state legislators to give women the right to vote" (Jenkins 2011, 133–4). This initial work was largely conducted in domestic spaces, such as parlors and in respectable locations such as churches and town halls. Early suffragists argued for the vote largely based on justice arguments: "women deserved political rights equal to those of men because, like men, they were citizens" (McCammon et al. 2001, 57; 2004). But the practice of American politics in the late nineteenth century was largely public, taking place "... in the male preserves of saloons, streets, and fields" (McGerr 1990, 867). This first wave had limited success: by 1900, the suffrage movement had only won four states in the country (Wyoming in 1869, Utah in 1870, Colorado in 1893, and Idaho in 1896).

Through the years of campaigning for the vote, there was not a singular suffrage movement but rather a number of affiliated organizations who advocated for women's voting rights. The National Woman Suffrage Association (NWSA) was headed by Elizabeth Cady Stanton and Susan B. Anthony. NWSA pushed for the federal enfranchisement of women but opposed the ratification of the 15th Amendment, which granted the vote to black men but not women (Jenkins 2011, 133). The American Woman Suffrage Association (AWSA) was headed by Lucy Stone and advocated for state suffrage and forms of partial suffrage, such as municipal or school suffrage, while supporting the ratification of the 15th Amendment (Wheeler 1995, 10). The 15th Amendment was adopted in 1870. These organizations merged in 1890 to form the National American Woman Suffrage Association (NAWSA) with Stanton, Anthony and Stone all in executive positions, and employing the state-by-state approach of the AWSA to work towards a federal amendment granting women the vote (Jenkins 2011, 134). But this period was marked by what they termed "the doldrums": the movement was making no measurable progress and its leadership was elderly and dying (Finnegan 1999, 115). In 1914, Alice Paul formed the Congressional Union, a younger, more radical movement for a national suffrage amendment drawing significantly on the British suffrage movement (Dumenil 2007, 22). I will focus my attention here on the National American Woman Suffrage Association (NAWSA) which employed a state by state approach and created and wielded the suffrage map.

After 1900, as new leaders stepped up, the movement began to move out of the "doldrums" and into a period of high activity. Part of this shift was tied to the rise of the "new woman": women were moving out of the domestic sphere and into domains traditionally male, shifting the social order between the sexes. As women were more visibly present in public – in stores and businesses, factories and offices, colleges and universities – the widely held assumptions of women's place in the home gradually weakened (McCammon et al. 2001, 53).

Another significant element was the suffragists

> ... inaugurated a period of stunning political experimentation that challenged the voluntarist hegemony.... Drawing on many sources, women created a period of rich possibility, adopting methods as innovative as anything they had attempted in the nineteenth century. The suffrage movement developed an eclectic mix of styles, partly inspired by both major political parties, socialists, trade unions, and English suffragists.
>
> McGerr 1990, 869[2]

This new approach involved a much higher public presence "... arguing visually and bodily for the vote through open-air speaking, automobile tours, balls, films, plays, tableaux vivants, parades, and other publicity generating stunts" (Madsen 2014, 288). And they shifted their argument to one of "expediency" – "emphasized that state policies increasing regulated the domestic sphere and that women could bring knowledge of the domestic sphere to the political arena in determining, for instance, how food, water, domestic violence, children's schooling, and even alcohol abuse should be regulated" (McCammon et al. 2001, 58). Such arguments framed suffrage in a way that did not challenge beliefs about women's roles but rather reinforced them in some ways, resonating with "widely held beliefs about women's appropriate place in society" (McCammon and Campbell 2001, 63). According to research conducted by McCammon and Campbell: "Suffragists' use of expediency arguments ... helped them win the vote" (McCammon and Campbell 2001, 74).

Part of this reinvigorated campaign was distributing literature and focusing on education, developing a rational, unemotional argument for suffrage, appealing to educated middle- and upper-class men (McGerr 1990, 870). But they also took to the streets for open-air rallies, parades, pageants. Such public displays were not just for winning over male voters, they also were empowering for women "giving participates a sense of pride, solidarity, and power" and was a means of claiming public space (McGerr 1990, 879). By the 1910s, ironically, men's political activism had largely "retreated indoors, to homes, nickelodeons, and saloons" (McGerr 1990, 879).

The first extensive suffrage campaign to use "many tactics of visibility, including advertising posters and signs; the use and sale of suffrage items such as stationary, lapel buttons, and baggage stickers; street speeches; and a storefront office dedicated to selling suffrage" was in California in 1911 (Sewell 2003, 85–6). Women as consumers had gained access to the downtown district, now they turned the tables and "borrowed the techniques of persuasion that had been used on them by downtown businesses" (Sewell 2003, 86). In a downtown such as San Francisco, there was "female-gendered shopping space" and "male-gendered office space" with sidewalks as a sort of "mixed gender space" (Sewell 2003, 89). Suffragists used the mixed space of the sidewalk to reach both men and women with their political speeches, targeting major intersections and major streets. The campaign in California had to move quickly: once they placed a suffrage referendum on the ballot, they had less than three months to convince male voters to support their cause (Sewell 2003, 90). They used the downtown landscape of San Francisco extensively to mobilize their forces, with speeches on major intersections along Market Street, setting up a storefront with well-decorated windows, bright with suffrage yellow, and distributing leaflets out front, and on Market Street "a large, permanent, electric sign" reading "Votes for Women" (Sewell 2003, 90–91). But it was more than the suffrage storefront: many stores featured suffrage banners and items in the suffrage yellow,

> ... shopwindows, from one end of the city to the other, blossomed in
> every known shade of yellow, and to point to the reason for the color,

copies of the prize poster, in dull olives and tan, lightened with yellow and flame, gave the campaign cry 'Votes for Women'.

College Equal Suffrage League of Northern California 1913, 56[3]

Shop-owners provided space for women's clubs and meetings, bringing women/consumers in while creating a positive atmosphere for the groups (Sewell 2003, 92).[4] The use of a single color, yellow, was a tactic borrowed from advertising, unifying their campaign with color and slogan ("Votes for Women").

It was in this reinvigorated campaign involving greater public visibility that the suffrage map was crafted and wielded.

"I made the map myself": the development and use of "that suffrage map"

No one should be deprived of the vote because of a change of residence,' said Winona Marlin. 'Look at New York State all in mourning,' pointing to the suffrage map with non-suffrage States black and suffrage States white. 'They say men want to paint the town red. We want to paint the United States white.

One-Minute Talks ... 1915

The origins of the suffrage map are far from straight-forward: its first appearance in 1908 was met with contention. In 1908, the primary magazine of NAWSA, *The Woman's Journal,* published a suffrage map on its front page (Figure 5.1). The brief statement published about the map explains it was a reprint of a map accompanying an article published in *Appleton's Magazine* by Bertha Knobe, concluding "It is instructive to see in how large a part of our country women now have some share of suffrage" (A Suffrage Map 1908). Just one week later, *The Woman's Journal* included a brief comment: Mrs. Catharine Waugh McCulloch writes:

The suffrage map published in The Woman's Journal of Jan. 11, and credited to Appleton's Magazine, is an exact copy (except that the four full suffrage States are made blacker) of the map prepared by Prof. Sophonisba Breckenridge and Miss Anna Nicholes in our municipal campaign in Chicago.[5]

Another clarification appears February 1st:

We are moved to comment on the surprise expressed here in Chicago over the item referred to. Rand & McNally (printers) got out our maps for us, and I can supply them to any who desire them. We owe it to Miss Anna Nicholes to see that she is credited with preparing such a map, independent of others who may have done so.[6]

The clarification is interesting in that, while giving credit to Anna Nicholes and acknowledging Rand & McNally, it also begins the distribution process with their offer to supply copies.

A month later, Knobe writes a lengthy rebuttal regarding the suffrage map to the editors of *The Woman's Journal*.[7] I will reprint it in its entirety because I believe Knobe raises some interesting points:

> I do not imagine 'that suffrage map' will bring undying fame to anybody, but I am moved to say that I have particularly enjoyed the amicable controversy over its authorship, inasmuch as I made the map myself, and sold it to Appleton's Magazine to illustrate my article on 'The Suffrage Uprising.' It was copyrighted, published, and so accredited by that magazine, and, at my suggestion, afterwards reproduced in the Woman's Journal, through the courtesy of the publishers.
>
> Precisely one month has been consumed in my effort to secure from Chicago the original map, with which, according to my friend and first accuser, Mrs. Catherine Waugh McCullough, my map is said to be identical, that I might have proof of my alleged plagiarism. It is enclosed for your inspection, and I think you will agree that a 'legal mind' is not needed to note the essential differences in makeup. The original map was made in Chicago at the time of the municipal suffrage campaign for municipal suffrage purposes only, and is not even complete as far as it goes. My map is altogether an enlargement of the idea, being an equal-suffrage map, with designations of the five kinds of suffrage obtaining in 28 States. Among States having school suffrage, I added Washington and Oklahoma, and Delaware to those having tax-paying suffrage, thus making the fourth State possessing these two forms of enfranchisement.
>
> As a matter of fact, I saw a copy of the original map in Chicago over one year ago, and it instantly gave me the idea for a number of suffrage maps, another one of which is to appear shortly in another magazine. It is not strange that my subconscious mind absorbed the general scheme of using dark and light spaces, latitudinal and longitudinal lines, in marking my map, for such markings are universally employed in map-making, whether they illustrate prohibition or the corn crop.
>
> In the 'school vote' brochure which the Woman's Journal issued later, along with my map, it states that the latter is incomplete, inasmuch as it does not include Florida, Mississippi, and Indiana. My map was compiled with the greatest care, after consulting with the leading suffrage authorities in the country, including one of the editors of the Journal, and various State officials. As a matter of accuracy, it would be valuable to the many suffragists who are appropriating the map for one purpose or another, and who care to correct it from time to time as the movement grows, to know what form of school suffrage exists in these three States, and when granted.
>
> As far as I am concerned, I hope everybody in the country will assiduously take to the making of suffrage maps, for it is a most effective way

to advertise the cause; but because that goodly company in Chicago happened to draw a municipal suffrage map of the United States in the year of our Lord 1906, it does not follow that every suffrage map published thereafter must be stamped as a plagiarism. So, solely as a matter of professional integrity—both on behalf of myself and my publishers—I am obliged to stand sponsor for 'that suffrage map.'

Bertha Damaris Knobe
New York City.
Knobe 1908a

Unfortunately, the suffrage map created in 1906 in Chicago as part of the campaign for municipal suffrage has not been located to date.[8] In additional to Knobe's references to it, several reports on the campaign for municipal suffrage refer to the map: "We have also published some maps showing the woman suffrage States and countries" (National American Woman Suffrage Association 1907, 69). There is no question it existed, there are only questions about what it actually depicted and its appearance.

Knobe's defense begins with a statement I have not seen before in working on women's use of maps in the Progressive Era: "I made the map myself." This is a very clear and proud statement but it also sets her map apart from the Chicago municipal suffrage map: no mapping company assisted with the production of this. In the *Appleton's Magazine* version, there is clearly the credit "Drawn by Bertha Damaris Knobe."

Knobe acknowledges she had seen the Chicago municipal suffrage map in 1907 but argues that the essential elements of such maps "are universally employed in map-making, whether they illustrate prohibition or the corn crop." It is kind of a "see one, seen them all" kind of argument, that the style of the Chicago municipal suffrage map was not unique but ubiquitous in the American landscape. It also suggests a familiarity with maps and mapmaking.

Knobe then goes on to describe how she enlarged upon the concept and "compiled with the greatest care," with input from "leading suffrage authorities." She even suggests the Chicago municipal suffrage map "is not even complete as far as it goes." Together, these statements suggest Knobe believes her map is more accurate and better than the other map.

Finally, Knobe boldly suggests her map would be useful "to the many suffragists who are appropriating the map for one purpose or another" and she hopes "everyone in the country will assiduously take to the making of suffrage maps" as a means of "advertising the cause."

In the letter, Knobe mentions the Chicago municipal suffrage map gave her ideas for "a number of suffrage maps, another one of which is to appear shortly in another magazine." That map appeared with the brief article "Votes for Women: An Object-Lesson" and is a world suffrage map (Knobe 1908b). While there may be questions about who is the creator of the first suffrage map of the United States, as far as I can tell, Knobe was the originator of the world suffrage map.

Whether it was Knobe's or Nicholes' map which became a suffrage icon, it is difficult to discern. Perhaps Knobe was correct in that a map is, in essence, a map and that it resembles many other maps, so is it plagiarism if a map of suffrage resembles other maps of suffrage when they are employing essentially the same basic map-making conventions? In fact, there may be earlier suffrage maps. In 1916 the Woman Suffrage Party of New York's *The Woman Voter*, published "The First Suffrage Map," dating from 1869 and the suffrage campaign in Wyoming.[9] Created by Hamilton Willcox, it is centered on Wyoming and includes only the edges of adjacent states, with Wyoming white and adjacent states in black. At the bottom of the page, it reads "The Map shows WYOMING in the LIGHT of FREEDOM; while the surrounding regions are in DARKNESS." Willcox was involved in the suffrage campaign in Wyoming, so it is possible he created this map as part of the campaign but I have not yet been able to locate an original copy. As a result, I do know how this map originally appeared, whether the text below the image is from the original or was added by *The Woman Voter*. Also dating from 1869, a *New York Times* article describes an Equal Rights Association meeting in New York City where a man named Barnes from Michigan stood and "commenced an incoherent rambling speech." At one point he "… unrolled in view of the audience a large map, having on its face several forms and figures, which he would explain, so that they might learn how to progress in the way of goodness, virtue, and reform." However, his views were not well received, he was heckled, and eventually "he was persuaded to take a back seat, not until he had declared, however, that the designation 'Equal Rites,' as applied to the Convention, was a decided misnomer" (Equal Rites 1869). No description is provided of the "map," but the article is subtitled "Moral Maps and Celestial Kites." These earlier versions of suffrage mapping may predate the suffrage map of 1908 but they do not appear to have had the impact on the movement that the 1908 map had. This may also reflect the changes in the suffrage movement in the early twentieth century, where the movement was employing different tactics involving a greater public presence, and the map became part of their new tactics of visibility.

The suffragists however seem to have been considering the use of a suffrage map before the Knobe map. The 1907 annual suffrage convention was held in Chicago in February. In the published proceedings, the minutes from the business portion of the meeting included: "Voted that the National Committee on Local Arrangements be instructed to prepare a suffrage map for use at the next National Convention" (National American Woman Suffrage Association 1907, 51). We are left to wonder if they had seen the suffrage map prepared by Nicholes for the municipal suffrage campaign: it seems quite possible as the major players in Chicago's municipal suffrage campaign were at the annual suffrage convention in Chicago and perhaps it was at the convention Knobe saw the Chicago suffrage map. In any event, they were certainly of the mind to visually representing the status of suffrage in the United States.

Regardless of who originated the suffrage map, once published in 1908, it took off as a persuasive image of the suffrage movement. In the Chicago suffragists' correspondence with *The Woman's Journal*, they offer to provide copies "I can supply them to any who desire them."[10] Almost immediately the National Suffrage Headquarters was advertising a leaflet based on the map at $0.02 a copy (New Leaflets 1908; Blackwell 1908).[11] In July 1908, a California suffragist Alice Park wrote in to offer directions on how women could create their own large version of the suffrage map to use at meetings and demonstrations, providing directions of where to order a blank map of the United States, have it mounted on cloth for display, and then colored or shaded. The end results:

> … with 'gumption' the details may be arranged, and a striking wall map evolved, at small expense.
>
> In case more States become free from week to week, the map may be amended by shading another State immediately after each favorable vote.
>
> The use of the wall map is obvious, both on the wall at Suffrage Headquarters, and as a portable object-lesson to display at meetings.
>
> Park 1908, 1

As a thematic map using simple shading to indicate categories, in the form termed choropleth maps, this form was easily used by both amateur and professional cartographers to visually display information and easily reproducible. The basic form had existed since 1826 when a French cartographer used the style to depict "the ratio of (male) children in school to the population of each department in France" (Crampton 2004, 43). In the case of the French map, the shading not only conveyed statistical information, it also captured a moral landscape: "a scale of moral values directly inspired the gradual shadings of the map. The shading gave the impression of light cast on the map, comparable to the light of knowledge" (Friendly and Palsky 2007, 240–241). A choropleth map also embodied the state-by-state approach of NAWSA, capturing the state-by-state progress of the movement, in which each state represented a markedly different battleground.

Maps such as the suffrage map invite readers to reflect of the status of their own state on the map, literally and metaphorically. Combining the moral implications of such mapping with the nature of viewers to reflect of their literal position on the map, results in a powerful visual argument. Mapping became part of this state-by-state approach beginning with the California campaign, who employed a slightly modified version of Figure 5.1 in their campaign.[12] Quickly, the suffrage map became a pillar of suffrage iconography, used in print and displayed publicly.

The suffrage map slowly evolved into propaganda. From 1908 to 1911, it merely colored the states that extended women suffrage, creating islands of black in a sea of white. However, the text framing the map edges the map towards propaganda, pointing out where women have the vote and where

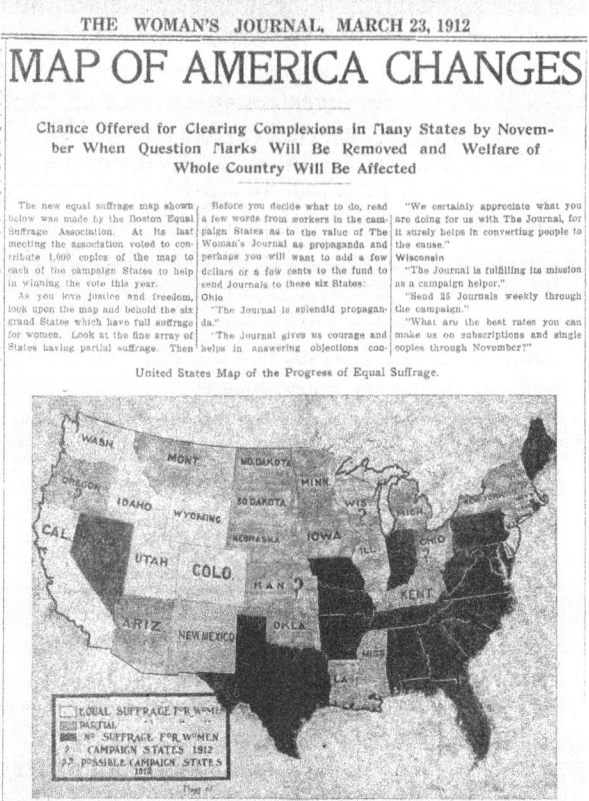

Figure 5.2 A 1912 example of the evolution of the suffrage map with the colors flipped: now states with suffrage are white and states without are black. From Agnes Ryan's "Map of America Changes" *The Woman's Journal* 43, 12: 91. By permission of the Wisconsin State Historical Society, Madison, WI (Image ID 130595).

they do not. The map used in 1908 in California offered a clear message to California voters: "California women have no votes."

The first step towards propaganda would not be taken until 1911, when the maps colors were reversed, with states with suffrage now being white and states without being black (Figure 5.2).[13] Several copies were published in 1911 and 1912 but had not yet reached propaganda status, lacking the persuasive rhetoric. In 1912, the modified map appeared again in *The Woman Citizen* but the rhetoric was beginning to take shape. The article surrounding the map invited readers to reflect on the map:

> As you love justice and freedom, look upon the map and behold the six grand States which have full suffrage for women.... Do you care? What

does it mean to you? Can you look upon the map unmoved? ... Look again at the map we are bent on changing. The questions marks stretch from coast to coast! If we win Oregon, Nevada, Kansas, Wisconsin, Ohio and New Hampshire, we have practically won the whole country to equal suffrage. Is it worth it? Shall we do it? Count your money, test your resourcefulness, consider what you can do. Look at the map of the United States as shown on this page and sing a new version of an old hymn thus:

> A flag to change I have,
>   A map to modify
> A never dying cause to save
>   Mankind to glorify.

Ryan 1912

Like the California map, the text invites the readers to consider their own state, literally and figuratively. By 1913, the suffrage map was being used extensively in the suffrage campaigns and being reproduced as posters and on other suffrage materials.

By switching from highlighting in black the states with suffrage to highlighting the states without, the map shifts from documenting success to identifying areas where work is needed. By doing so, the map offers a visual form of the rhetorical argument that the vote emancipates women while denying women the vote keeps women shackled and "in the dark." Moral maps had long employed a color scheme equating white with purity, virtue, freedom and black with uncleanness, ignorance, slavery.[14] The landscape of suffrage was actually quite complicated, ranging from no suffrage at all through forms of suffrage (such as municipal or school suffrage) to equal suffrage (see Figures 5.1 and 5.2). But for some maps, the various categories were boiled down to just black and white, perhaps due to the medium, the production methods, and/or budgeting concerns.

The shift in colors was very much tied to an effort to have the map reflect suffrage ideology. States without suffrage, especially those surrounded by suffrage, came to be referred to a 'black spot' on the map, such as "The suffrage map showing Nevada as the last 'black spot' in the West was printed in every newspaper and on every leaflet, put up in public places and on large banners hung in the street" (Harper 1922a, 398).

Another oft used set of slogans with the suffrage map was: "Votes for Women a Success. The Map Proves It. Would any of these States have adopted EQUAL SUFFRAGE if it had been a failure just across the Border? Imitation Is The Sincerest Flattery!" This image/argument, used throughout the country, creates the impression that the diffusion of woman's suffrage is natural, flowing effortlessly from West to East. The point is hammered home with the text, lest the viewers not get the visual message. The text proclaims what the map alone cannot. McCammon et al argue in their analysis of gendered opportunities and successes in the women's suffrage movement that one of the opportunities leading to suffrage success was the success of suffrage in a state or neighboring state: "As the public witnessed women voting in minor elections locally or in major elections in neighboring states with competence and good results, views towards women's political participation liberalized and acceptance of suffrage rights grew" (McCammon et al. 2001, 54). McCammon and Campbell further suggest "As attitudes toward women and the vote shifted in one state, it appears that they influenced attitudes in neighboring states, leading neighboring states as well to grant suffrage" (McCammon and Campbell 2001, 76). Suffragists appeared aware of this, as they extensively used this rhetorical device with the map across the country.

As the movement gained momentum, the National American Woman Suffrage Association formed a company to produce and distribute suffrage materials.[15] In addition to leaflets and pamphlets, the National Woman Suffrage Publishing Company, Inc., produced books, a "National Suffrage Library" in several volumes, as well as posters, maps, playing cards, stationary, calendars, postcards, and parasols: "It has proven so successful as a business proposition that in January of this year, after two years of work, it declared a dividend of 3 percent" (Stapler 1917; Suffragists' Machine Perfected … 1917).[16] Suffragists put the suffrage map on a wide-variety of mediums, getting their message out wherever they could. This included: fliers, posters and banners, postcards, hand fans at sporting events, calendars and drinking glasses as well as copies in newspapers and magazines.[17] These promotional items were distributed to male voters – such as the calendar distributed to street railway workers and male parade marchers, and a suffrage map tucked into baseball programs.[18] In the "Empire State Campaign of 1915,"

> Suffragists covered the visual landscape with advertisements. They distributed 149,533 posters – 'thousands' hung from trees, on fences, and in the windows of houses, apartments, and storefronts. Other posters decorated the interiors of banks, moving picture and vaudeville theaters, and other businesses.
>
> Finnegan 1999, 61–63

Changes in the suffrage map, while a sign of the success of the movement, proved to be a headache for the Publishing Company:

> 'Minnesota has won presidential suffrage!' was the shout that rang through the ballroom of the Statler, where preparations were on for the Jubilee Convention today.
>   Maine on Wednesday, Minnesota on Friday!
>   Then the horrifying thought – those maps!
>   The swiftest runner in the world could not keep up with the suffrage map. One of the prize 'layouts' in the Woman Citizen's Jubilee Convention number is a two-page spread of the suffrage map, brought

right up to the minute, so the editors thought. Now it is all wrong—and no one is the least bit sorry.

The Jubilant Diary … 1919, 924

And from Ida Harper's *History of Woman's Suffrage*, published soon after the passage of the Nineteenth Amendment:

> … the company was 'bankrupted' trying to supply 'suffrage maps' up to date, for as soon as a lot was published another State would give Presidential or Municipal suffrage and then the demand would come for maps with the new State 'white,' and thousands of others would have to be 'scrapped.'

Harper 1922a, 531–2

While keeping the suffrage maps up-to-date proved challenging, it represented hard-won success, for each up-dated map represented a step closer to their goal of suffrage for American women.

The suffrage map proved to be an effective rhetorical device in their campaign for the vote. As a means of arguing for the vote, the suffrage map was practical: "… suited to action, movement, and the management of space in real time" (Jacob 2006, 81). In the quickly evolving context of political activism, the map is mobilized, made to be taken to the streets, to be handed out, displayed, and explained at/in the moment (Figure 5.3). It is not a "precious" map, with only one in existence, rather it is moveable, disposable, capable of being crafted quickly and on the fly. Using a simple style that is easily reproducible and easily adaptable to different mediums, results in an image both fixed and easily moveable across space (Latour 1987). NAWSA's production of suffrage materials was important to unifying and motivating Americans, serving as an anchor: "It provides geographical discourse with a referent by anchoring it in a visible reality. Without the map, this discourse is purely ideal, the object – formless and indefinite – of a postulation rather than of knowledge" (Jacob 2006, 30). But the map was not only effective as an immutable mobile, but also as a form of propaganda: "Propaganda aims at persuading large groups of people to believe something or act in a way that they would not, in the normal course of events" (Pickles 1992, 201). Maps have long been used by those in power, masculine hegemony, to influence and control populations. Ironically, American women were now attempting to turn the tables and use the power of maps on American men.

In employing a map of the United States, suffragists were appealing to not only logic by using the tools of social science, but they were also appealing to a sense of nationalism. The outline of the United States is iconic, serving as a shorthand for the country. Maps have long been used as parts of national movements, with public displays of maps used to invoke patriotic feelings and mobilize citizens (Cairo 2006, 374). Suffragists put the map

Figure 5.3 The suffrage map in action. A suffragist holding a suffrage map up at an open-air meeting outside a factory in Pennsylvania, Historical Society of Pennsylvania (League of Women Voters Records [Coll. 2095]).

in the windows of their offices, they included it in their parades, they even painted it on a building opposite of the site of the 1916 Republican Party Nominating Convention (Finnegan 1999, 62). But they took it a step further: suffragists lobbied and successful placed the map in the halls of government. In West Virginia, the suffrage map was placed in the lobby of the state capital (Harper 1922a, 688). In New Hampshire, "A large illuminated 'suffrage map' was framed and put in the State House and other public places" (Harper 1922a, 405). In Virginia, "Mrs. L. S. Foster, president of the Williamsburg Equal Suffrage league had placed on the walls of the old courthouse there, built in 1769, the new suffrage map, starred for the states which have ratified the Federal Suffrage Amendment. As the list increases, Mrs. Foster will add the stars" (Virginia Women ... 1919, 266). Capitals, whether state or national, are the halls of power and are the proper place of maps, where maps belong. So too are the collections of geographical societies: NAWSA donated suffrage maps to the American Geographical Society's library.[19] By placing their maps in these contexts, the suffragists were making the political argument that their maps are legitimate depictions of reality and should be taken seriously while appealing to sense of nationalism.

In addition to using national maps, a handful of states created regional suffrage maps. Oregon created at least one regional suffrage map as part of their campaign (Figure 5.4). Oklahoma also created a regional suffrage map with the slogan "Make Oklahoma White" in their 1918 campaign.[20] And a men's group supporting suffrage created a Pennsylvania map, breaking down suffrage status county by county.[21] In these examples, the maps are centered on the state and include at most the margins of adjacent states and their suffrage status.

In addition to employing the suffrage map as part of their rhetorical campaign, state suffrage organizations used maps in a variety of ways in their efforts to pass suffrage in their states. In their account of the successful California campaign, a suffragist explains how they carried out an extensive automobile campaign in what they called the "Blue Liner," particularly in the Oakland and Berkeley areas:

> There is a loose-leaf filing-book beside me as I write, a foot square and very fat. On the front page is pasted a road map of Contra Costa, trailed over and over with pencil marks, showing routes. Inside are filed some ten dozens of letters of diplomatic correspondence, written, week by week, in the continuous effort to open wide the gates of about twenty little towns of Contra Costa to the entry of the Blue Liner.
> College Equal Suffrage League of Northern California 1913

It seems this map was used both for navigation and for strategizing about reaching those "twenty little towns." An account of the New York campaign describes how "Big Boss" Mary Garrett Hay, the Greater New York leader, used maps: "On the wall of her office are five carefully detailed political maps

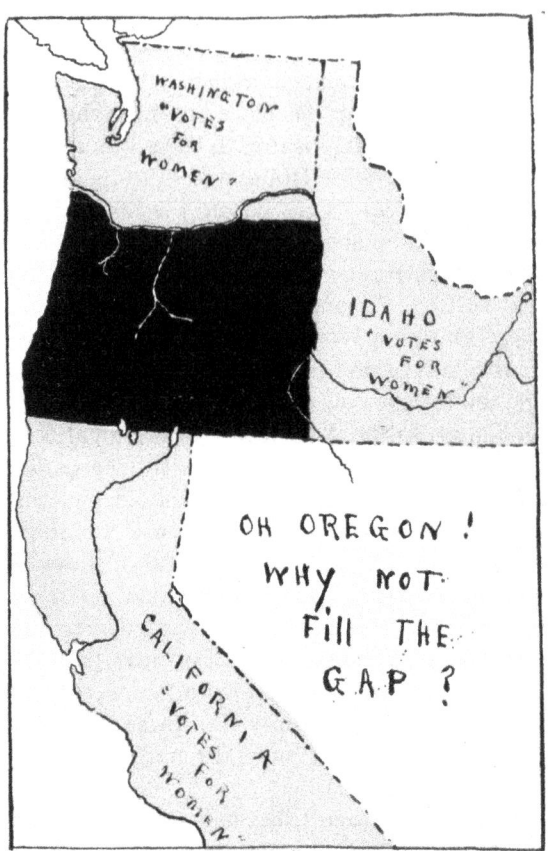

Figure 5.4 Example of a regional suffrage map: Oregon map from the Susan B. Anthony Ephemera Collection, Huntington Library, San Marino, CA (ephSBA Postcards volume 2, #27).

of the five boroughs, where points of difficulty may be pointed out in consultations" ('Suffragists' Machine Perfected ... 1917). The maps used in the Sixth Campaign District (south central New York State) are published maps of the region with sketching over the map, dividing the district up as they planned their work.[22] As the women strategized about advancing their campaign, they plotted over maps, not unlike a general in a battle, planning their attack.

NAWSA also published additional maps towards the suffrage effort. A report in the Proceedings of the Forty-Sixth Annual Convention of NAWSA (1914) describes their efforts to establish a "Woman's Independence Day" on the first Saturday of May. While President Wilson declined to declare the day a national legal holiday, many women in the US celebrated the day. Suffrage leaders documented the ways women celebrated the day and created a map: "the map sent out from the Chicago office showing a dot for every city and town where a meeting was held made the United States look as if it had been bountifully showered with pepper. Not a state in the Union was silent, not even the suffrage states, and many added parades and other events to the regular program...." (National American Woman Suffrage Association 1914, 43). Unfortunately, this map has not been located to date. At the Jubilee Convention in 1919, a report was given by Mary Garrett Hay on fund-raising efforts entitled "Make the Map White," where states who met their fundraising quotas received a "red star": rather than the suffrage slogan of "making the map white," Hay "really wanted to make it red with stars" (NAWSA 1919, 46). Ultimately, thirty-two states received the red star. This map has also not been located. Similarly, the Oklahoma suffrage committee reported they mapped out where local suffrage committees were organized in their state: "a map of Oklahoma was shown almost obliterated with tiny red stars. They were told 'Every one of these represents a suffrage committee organized since January'" (NAWSA 1919, 102). This map also has not been located. Besides propaganda and mapping strategies, the suffragists were using the maps to indicate to their constituents the success they were having in mobilizing their forces. The use of maps for a variety of functions by the suffrage movement suggests a deep familiarity with maps and their various uses.

Over time, the suffrage map became so pervasive in the American landscape that the suffragists would refer to it, even if it was not there. Crystal Benedict, addressing the suffragists' approach to politicians stated in 1914:

> Our plea is simply that you look at the little suffrage map. That triumphant, threatening army of white States crowding rapidly eastward toward the center of population is the sum and substance of our argument. It represents 4,000,000 women voters. Do you want to put yourself in the very delicate position of going to those women next fall for endorsement and re-election after having refused to report a woman suffrage amendment out of committee for discussion on the floor of the House?
>
> Harper 1922a, 429

In an essay entitled "Colored Women as Voters," by Adella Hunt Logan, published in the NAACP's *The Crisis*, there is no map but Logan refers to it: "The suffrage map shows that six states have equal political rights for women and men, and that a much larger number have granted partial suffrage to women" (Logan 1912). While black women numbered among the suffragists, the question of race was a significant one for the movement and the map.

The 'dark spot' on the suffrage map: race, the map and the movement

> A map of the United States, showing the suffrage states in white, the no suffrage states in black, and the partial suffrage states with dotted or shaded lines, reveals the southeast seaboard and its hinterland in solid black.

This was the 'solid south.' It was the land of slavery. It is the land where women have no vote upon any questions or candidate. It is the land of chivalry. It is the land of child labor.

On the suffrage map it is our darkest America. Missionaries are needed to work in the south.

The Black Belt 1914

The roots of the suffrage movement are deeply intertwined with those of the anti-slavery movement and temperance movements in the 1830s and 1840s:

The nation was seething with the discussion of temperance and slavery... . Women were as vitally interested in these questions as were men and they could no longer remain silent and inactive. Their first timid attempts at speaking in public and taking part in conventions were made in the temperance movement and here they met with a determined attempt by the men to suppress them.

Woman Suffrage 1920, 447

This connection between women's rights and abolition was multi-leveled: there was the advocacy for basic human rights and justice for all, but there was also a geographical element to both. Certainly, for those Black Americans enslaved in the American South, there was a geography of slavery and a geography of freedom, depending on their spatial location. Spatial information to move from the South to the North was essential to transcend this geography. On a figurative level, Martha Schoolman argues for "abolitionist geographies" in her book by the same title (2014). Schoolman examines abolitionist literary texts from between 1834–1859 and their promotion of "geography as a key discourse of abolitionist political intervention" (Schoolman 2014, 1). In their texts, American abolitionists, as they spun their narratives and sought to shift their readers' mindsets on slavery, were critically remapping the landscape, beyond just North and South but also transcending US boundaries.

There was a distinct tradition of abolitionist mapping. Factual maps such as the 1860 census map of the population density of slavery in the American South and Sidney Morse's 1863 map in *A Geographical, Statistical and Ethical View of the American Slaveholder's Rebellion* were used to draw Americans' eyes to the isolated Southern practice of slavery (Schulten 2012, 119–155). Metaphorical maps, such as such as Thomas Clarkson's abolition map (1808) were to be used to visually map the abolitionist network, creating a genealogy of abolitionists in the form of a hydrologic map (Wood 1999). Finally, propaganda maps like the 1847 "Moral Map of the United States" were created and distributed to mobilize allies, depicting a United States where the slaves/slavery are coded black and concentrated in the South and the "free" north is white.[23] An 1854 version ramps up the rhetoric and add color and decorations, with the South now being portrayed as blood red and on the left margin of the map, a black man is bound by a rope flowing from the American flag at the top of the map.

Whether it is text or an image, "black" and "white" are tied to complex metaphors as well as into the racial landscape of early twentieth century America. The association of white with light and good and black with darkness and evil have roots back at least to the Christian Bible (Dalal 2002, 140–4). As time passed, in European cultures, white became an indicator of cleanliness, favor and honesty or legitimacy, while black indicated dirt, immorality or illegality, evil, and death (153–7). The first choropleth maps used black and white to convey the "... unenlightened and enlightened regions of the country," creating a moral cartography (Crampton 2004, 43). An interest in "moral statistics," statistics on crime, literacy, suicide, resulted in more moral maps through the 1860s (Robinson 1982, 156–70). Eventually, this moral cartography was used in the classic colonial discourse of bringing culture and civilization to "even the darkest and most barren reaches of the empire," terms which referred in particular to the African continent (Minor 1999, 149). British and American explorers and missionaries were bringing the light of science, democracy, Christianity, and capitalism to a place associated with 'darkness' on many levels – skin, savagery, paganism, wildness (Jarosz 1992, 106–7). Educated American women would have been aware of the use of this metaphor from the accounts of explorers, such as Henry Stanley's *In Darkest Africa* (1878) and *Through the Dark Continent* (1890), as well as from popular fiction, such as the works of H. Rider Haggard (Brantlinger 1985).

American women may also have been familiar with the maps, such as the 1847 "moral map" which "Northern abolitionists had used to illustrate the white moral purity of the free states compared to the black evil of the Southern slave states" (Reynolds 2005, 40). They may have been aware of social/ moral mapping projects such as those of Charles Booth and William Stead. Stead's maps used blue and black to represent "lowest class" and "casual earnings," which were associated with criminality, roughness, and disorderliness (Kimball 2006, 370–1). Women involved in social settlements were aware of and influenced by Booth's maps (Residents of Hull House 1895, 57). Stead visited Chicago and was a frequent late-night visitor to Hull-House (Churchill 2002, 10; Downey 1987, 155). Stead's *If Christ Came to Chicago!* (1894) featured a map of a block in Chicago's First Ward, using black, grey, and red to indicate the locations of saloons, pawnbrokers, and brothels, juxtaposed with an image of Christ ejecting the money changers from the temple. Booth's and Stead's maps are essentially large-scale choropleth maps where the units mapped are city blocks (Booth) and buildings (Stead), with a moral element conveyed using color.

While the suffrage movement had long ties to the abolition movement, the subject of black voting rights had divided suffragists since the Fifteenth Amendment enfranchised black male voters in 1866 (Buechler 1986, 5–6). Black women believed, just as white women did, they needed the ballot to help themselves and their communities (Terborg-Penn 1998, 79). White southerners were concerned about black voting in general, some fretting the "moral superiority" of black women would result not only in their voting but also in their resisting overtures to having their votes bought. For the most part, black women were excluded from the pre-dominantly white suffrage organizations (Dumenil 2007, 22). In the name of political expediency, suffragists were turning their backs on former allies, compromising on their approach (moving from equality of the sexes to the vote being necessary for "women's work"), and in general, using all the tools in the toolbox, including racial rhetoric, to advance their cause.

Playing on this use of black/white, backward/enlightened, dirty/clean, was suffrage rhetoric about "cleaning," part of women's domestic duties. Much anti-suffrage rhetoric was based on the notion of separate spheres: men's place was in public, women's in the home. It was feared if women were given the vote, they would become "manly," while men would become feminized, a frequent topic for anti-suffrage editorial cartoons (Buechler 1986; Palczewski 2005). Domestic cleaning was part of "women's work," and the use of the brooms emphasized that granting women the vote was not about women stepping out of their place but was instead about the ramifications of political decisions on the home (municipal housekeeping) (Buechler 1986, 27 and 166). Women in suffrage parades often marched with brooms tied with yellow ribbons (Big Suffrage Party 1915; Sewell 2003). Proclaiming "A Clean Sweep for Suffrage," the brooms transformed the streets from public space to domestic space. The rhetoric of removing "offensive black spots," then, played on this domestic imagery: as good "housekeepers," the suffragists would of course want to make the nation "spotless."

Another play on black and white combines the color metaphors, the racial issues and cleanliness. Broadsides used in many state campaigns ask the question: "Won't you help us make [state] white?" This provocative statement refers to the map's colors, as well as to racial concerns over African American women voting, but also to cleanliness/housekeeping. An account of the Nevada suffrage campaign describes how "the suffrage map showing Nevada as the last 'black spot' in the West was printed in every newspaper and on every leaflet, put in public places and on large banners hung in the streets" (Harper 1922b, 398). *The Suffragist* reported "women all over the country desire that the offensive black spot be removed from the center of the white field on the suffrage map," incorporating domestic metaphors of cleanliness into the rhetoric (Vernon 1914, 6). Nevada passed the suffrage amendment, becoming "white" in 1914. Missouri and Georgia also employed this racial rhetoric (Harper 1922b, 343). In Georgia in particular, it reads as a racial statement: "Georgia's complexion on the suffrage map has been pot black" (Georgia vs. Connecticut 1919). Southern suffragists argued that giving the vote to white women could help preserve white supremacy ("The Movement Comes of Age" 2005). But it was not just southern states using this rhetoric. In addition to Nevada, a Michigan newspaper printed the suffrage map in 1918 with the headline "Make Michigan 100% White on the Suffrage Map" (Make Michigan … 1918). The origins of the phrase "black spot" have not been established, but as the color black became associated with dirty, illegal, immoral, a "black spot" on a map would be seen to indicate "bad" places (Dalal 2002, 157–60). The map and the text play on the classic dichotomy of white as good, black as bad, but also refer to the color of slaves. This appears to have been employed in many anti-slavery maps, such as the 1847 "Moral Map of the United States" and the 1860 census map of the population density of slavery in the American South, where the slaves/ slavery are coded black and concentrated in the South and the "free" north is white (Schulten 2012, 119–155). In some ways, this was necessary given the printing technologies of the day: almost everything was reduced to black and white. But it is their use of black and white and the rhetoric shifting the maps from relatively straightforward facts into propaganda.

The linkages between suffrage and the abolition movement are kept at the forefront of the public's attention through the suffragists' use of a quote from President Abraham Lincoln's landmark "House Divided" speech. In a New York newspaper from 1914, the text accompanying the suffrage map paraphrases President Lincoln: "'This Government cannot exist on half slave and one half free'" before continuing, "Neither can the country continue to be governed one half by half the people, the other half by all the people" (The Woman Suffrage Map of 1914). States where women did not have the vote were therefore assumed to be "slave states." Further emphasizing the parallels between slavery and suffrage, suffragists used women to represent the suffrage status of each state in parades and pageants. Women who represented "no suffrage" states would be dressed in black and be in shackles: "Symbols of bondage and servitude, they brought the meaning of disfranchisement to a personal, material level that associated women with the lowest members of a society: slaves, criminals, traitors" (Finnegan 1999, 84). The Civil War and the abolition movement were still in public memory. The suffragists' use of the map ties nationalism and identity to their political position, capturing the "divide" between suffrage and non-suffrage states, creating or recreating an East-West divide for a population that still remembers a North-South divide. As a result, the suffrage map represents a call for unity.

While theoretically black men had received the vote with the 15th Amendment and black women with the 19th, particularly in the South disenfranchisement methods were used to prevent them from exercising their rights, ranging from threats of harm if they voted to tests on their ability to read and interpret the Constitution. For some Black Americans, they did not truly receive the right to vote until the 1960s Civil Rights movement.

"Living suffrage maps": embodying the suffrage map

> An idea that is driven home to the mind through the eye produces a more striking and lasting impression than any that goes through the ear.
> Tinnin 1913

Glenna Smith Tinnin, in the above quote, was discussing pageants but it also captures the power of the suffrage map: together they were part of the campaign to win women the vote. Maps, parades, and pageants were components of the invigorated suffrage campaign, drawing from the British suffrage movement's use of public spectacle. As the suffragists took to the streets, they took the suffrage map with them in a variety of forms. Some forms were more conventional, such as carrying the suffrage map or creating a float with the map on it. Other forms involved women becoming the map, with women representing each state and their appearance indicating its status. These "living suffrage maps" represent a form of performance cartography.

The study of maps often focuses on their knowledge, their production, their technology, their design. More recently studies have considered the ways in which they are utilized, considering both their practice (how they are used) and cartographic culture (the ways in which a society understands and employs maps), sometimes using the "performative" to consider how maps are both created and consumed (Perkins 2009, 127). However, there is also the concept of "performance cartography." Traditionally, in Western culture, maps are conceptualized as material artifacts communicating spatial information. Woodward and Lewis have suggested there are in fact three interrelated mapping traditions: in addition to a material tradition, there is also a cognitive map tradition and a performance tradition. Performance cartography is strongly embedded in cultural practices and may involve mapping in rituals or performances through gestures, rituals, songs, dances and other symbolic acts/movements (Woodward and Lewis 1998, 3). Through these actions, humans convey their conceptualization of real and imagined places as well as a sense of spatial power and control related to territoriality. The performance of maps and mappings involves a more pronounced narrative and storytelling element as it conveys spatial relations (Chávez 2009, 168). Yet all storytelling involves a collaboration between the teller and the listener/audience, as does performance. In involving maps in suffrage events such as parades and pageants, suffragists are publicly communicating their worldview while engaging with their audiences to share this view.

Parades were a popular public display of unity and power in the nineteenth century America, largely composed of men who marched with their comrades "to express a common social identity" and to arouse attention and support for their cause (Borda 2002, 28 and 31). Prior to suffrage, women were active in popular politics in traditional feminine support roles, attending meetings and supporting parades by making food, sewing banners, decorating meeting halls (McGerr 1990, 867). Largely relegated to supporting roles, some women did more actively participate:

> ... younger women and girls sometimes participated in campaign events, mainly as symbols of virtue and beauty. Middle-class girls dressed up as the Goddess of Liberty and stood on the lawns of their homes as parades marched past. Young women presented banners to marching

companies. For parades, particularly in the Midwest, women representing Liberty or individual states rode in wagons or floats.

McGerr 1990, 867

When the suffrage movement began to employ parades in the campaign,
they played on two elements: the public display of unity and power AND the
representation of democratic ideals.

The first suffrage parade was held in New York City in 1908. Denied a
parade permit and composed of only twenty-three women, their parade
drew a crowd estimated to be over a thousand men (Borda 2002, 31 and
49; DuBois 1995, 239). Parades presented suffrage as a movement unifying
women across socio-economic levels – factory girls, married women with
children, university professors, society ladies – presenting the women as a
serious lobbying block (Jenkins 2011, 136). Most parade organizers encouraged participants to dress in white, further unifying them by color (dress
and skin) while using the color's association with purity and virtue (Jenkins
2011, 139). Within the parade, they did not march randomly but were organized into like groups, with each group representing a different argument
for suffrage, such as housewives and mothers emphasizing the struggle to
protect and care for their families and women from different countries signi-
fying the global reach of suffrage (Finnegan 1999, 90–91). Women marched
with banners reading "Votes for Women," "There is a word that is dearer
than Home or Mother or Country: that is Liberty," and "All This Is The
Natural Consequence of Teaching Girls to Read" (Borda 2002, 34).

Parades took place on major thoroughfares and were well-organized displays of suffrage strength, with women from the surrounding area coming
in for the event. The New York City suffrage parade of May 1912 reportedly
had 10,000 women participating. A November 1912 parade in New York
drew 10,000 participants and reportedly nearly 400,000 onlookers (Stovall
2013, 19). In the case of the Connecticut suffrage parade of 1914, suffragists
made their way from the Connecticut Woman Suffrage Association headquarters to the state capitol, "where women attended hearings on woman
suffrage at the state House of Representatives" (Jenkins 2011, 136). They
were not just claiming male public-space of the street but symbolically occupying the spaces associated with political power.

This was taken to the highest level when a national suffrage parade was
held in Washington DC in 1913, on the day before Woodrow Wilson's inau-
guration and following its procession route (Jenkins 2011, 140). The procession was orchestrated by Alice Paul, who

> envisioned an event that would arrest and impress viewers with a kaleido-
> scope of color, people, and ideas. Grace, movement, and feminine beauty
> would be the tools that would lift the idea of suffrage onto a higher plane
> of public debate than it had ever attained in the United States.
>
> Stovall 2013, xi[24]

Suffragists across the country were mobilized: hikers set out from New York and newspapers followed their progress as they made their way to Washington; a splendid pageant was planned for the steps of the Treasury; and thousands of women came from across the country. It is estimated there were at least five thousand marchers, with some estimates suggesting as many as eight thousand, and the marchers organized into sections with a herald announcing each section (Stovall 2013, 31). Newspapers estimated the parade audience to be between 300,000 and 500,000 people, predominantly men (Chapman 1999, 344).

The suffrage map was integrated into the parades in a variety of ways. In the 1913 Washington "Woman Suffrage Procession," the suffrage map appeared, I believe, three times – as a float, as a "living suffrage map," and carried by Lavinia Dock. Accounts describing the parade plans included "… the banner float, with a map showing nine States of light and thirty-nine of darkness … the last float will be a State car with 9 women dressed in white representing 'light' and others walking around dressed in black, representing the 39 States which have no suffrage. This float will carry a banner with the words of Lincoln, 'No country can exist half slave and half free'" (Parade to Glow … 1913).[25] This description is deeply ironic, considering black women were marginalized in the parade, shunted to the end and not allowed to participate in the pageantry (Frost-Knappman and Cullen-Dupont 2005, 296; Stovall 2013, 24). One rather poor photograph of the map float exists from the Washington parade, capturing a horse-drawn float with a "billboard" of the suffrage map with the slogan "9 States of Light Among 39 of Darkness."[26] Lavinia Dock, a nurse and "Surgeon-General" of the Albany pilgrimage, marched at least part of the way from New York to Washington with her "walking suffrage map" around her neck (at least one photograph of the hikers captures Dock with her map).[27] In a variety of parades, "living suffrage maps" or "human suffrage maps" were employed, as marchers, on horseback, or on floats. A report on the South Dakota suffrage campaign explains how in a Sioux Falls parade, "Mrs. Lewis Leavitt had a human suffrage map (young girls), followed by a float with women dressed in national costumes of the full suffrage countries and other features" (McMahon 1918, 508). A parade held in Washington in 1915, featured "a group of mounted women, representing the eleven States and Alaska, in all of which women are enfranchised; [and] another group, representing the thirty-seven unenfranchised states" (McGerr 1990, 877). In addition to living suffrage maps were often women dressed to represent Liberty or Justice and even America herself.

Sometimes in parades, sometimes just on their own as a "soapbox speaker," suffragists put on the map. Dock wrote in to *The Woman's Journal* in 1913 and described her map and her approach:

> A new and satisfactory method of propaganda is my walking suffrage map. I got an outline map of the U.S.A. from the Survey Bureau in Washington. It is about 24 × 16 inches, I colored it in water colors in

strong, striking tints—full suffrage States, rich blue; partial suffrage, pink; non-suffrage, brown. I wear it sandwich fashion, and walk about my crowded streets. It attracts everyone's eye, and an explanation of the colors excites deep interest and makes a great impression. Men are much impressed by the ocular proof of our advance, and after little talks in groups of three to ten, many sign slips. The colored map is, I think, very valuable, as many people receive impression more strongly through the eye than the ear.

Dock 1913

Dock encourages other suffragists to use the map to engage the public and make the case for suffrage.

But suffrage parades were not all love and roses. In particular, the 1913 Washington D.C. parade resulted in suffragists being both verbally and physically assaulted. Parade organizers had a parade permit and had attempted to coordinate with the District of Columbia police but found them reticent. The 5,000–8,000 marchers were overwhelmed by the estimated 500,000 audience to the point of stopping the parade at times. Ultimately:

Women had to fight off throngs of spectators along the entire parade route, some of whom attempted to trip the women and climb on their floats. The parade ended hastily as heckling spectators in the sidelines suddenly evolved into an angry mob that broke out onto Pennsylvania Avenue and completely swarmed the marchers. Hundreds of participants were injured and the U.S. Cavalry was called in to restore order.

Borda 2002, 46

The chaos at the suffrage march resulted in Congressional hearings on why this violence occurred, especially as Wilson's inaugural parade the following day was peaceful (Bernstein 2005, 72). D.C. police had not only not intervened but behaved at times antagonistically towards the suffragists. Ultimately, the chaos, violence, and subsequent hearings were widely covered by the press, bring more attention to the cause.

Pageants were another medium suffragists used to draw attention to their cause and to mobilize voters: "The artistic and allegorical beauty of the floats and dances will appeal to the spectator's eye; the arguments on the banners and transparencies will appeal to his reason and sense of justice; but above all, the ocular proof showing how far the advance of equal suffrage has actually gone will appeal to his political common sense" (Blackwell 1913a). Some of these pageants involved "living suffrage maps." Pageants were an extremely popular American outdoor rite in the early twentieth century, requiring the participation of huge numbers of amateur performers in an event staged around a theme, usually related to social reform (Blair 1994, 118). They were a means for the suffragists to "set forth our ideals and aspirations more graphically than in any other way" (Moore 1997, 89–90).

Writing before the 1913 Washington parade and pageant, the pageant chair, Glenna Smith Tinnin, suggested: "The pageant can show that the extent of this women's movement is world-wide. It can present pictorially the countries that have equal suffrage, the countries in which women have municipal suffrage, and those where bills for woman suffrage are now before the Legislature" (Tinnin 1913, 50). Not all pageants included a living suffrage map, but there is evidence of the map being included in several pageants. In a 1916 pageant in St. Louis, women representing states with full suffrage are dressed in white and have shields, those representing partial suffrage are in grey, while those representing states with no suffrage are in black and shackled. In album/scrapbooks of Carrie Chapman Catt documenting the suffrage movement are several photographs depicting different staged tableaus of "Enfranchised and Unenfranchised states."[28] In 1916, a plan to have a tableau at an event in New York City went awry when their permit was rescinded when "it was pointed out that the tableaux, showing what States had and what States had not woman suffrage, was in the nature of a political propaganda" (Suffragists Silent ... 1916). Living suffrage maps brought to life the suffrage map, putting into motion the map used across the country in a highly symbolic context.

Parades and pageants were a way for women to claim male-dominated public space (Jenkins 2011, 136). By in-mass occupation of streets and roads, suffragists brought greater visibility to their place in a sphere they already occupied in many ways. Through parades and other public spectacles, "They were not temporarily passing through the public sphere but were *in it* in order to be *of it*." (Schultz 2010, 1135).[29] And in their own parades and pageants, women "... were reclaiming the right to determine whether and how they would be used as symbols" (McGerr 1990, 877).

Women representing states and ideals has precedent, with a traditional cartographic trope of women personifying continents on the title pages of atlases or in the cartouches, as illustrated by Brian Harley in his essay "Maps, Knowledge, Power," using the title page of Ortelius' *Theatrum Orbis Terrarum* (1573) (Harley 1988). In this image, the continents are personified as women, with only Europe fully clothed. In addition to continents, women were used on maps to personify ideals such as victory, justice, or liberty, usually portrayed as European women (see Lady Liberty in Figure 5.5). There was also a sixteenth-century European tradition of parades in which the king, duke or princess rode into the city with an immense entourage, with some costumed as allegories, such as continents (Le Corbeiller 1961, 209). Harley suggests female sexuality on maps was "often explicit for the benefit of male-dominated European societies," objectifying women for the sake of the patriarchal gaze (Harley 1988, 76). Whether continents or virtues, these women can be interpreted as either in need of men's security/protection or as a body to be "taken," and both interpretations assume women's vulnerability (Petto 2009, 68). The objectification of women on maps certainly creates a conundrum for women map viewers in earlier historic periods.

Figure 5.5 "The Awakening," by American illustrator Henry Mayer in *Puck* magazine's suffrage issue (1915). Employing the white and black imagery of the suffrage map, Mayer has added to the black, the bodies of women struggling, reaching their arms out to Liberty as she strides across the country with torch of enlightenment. Library of Congress (Illus. in AP101.P7 1915 Case X [P&P]).

They were not part of the perceived audience for maps, rather they were part of the decorations: adding visual appeal and possibly rhetorical punch but not contributing to the knowledge/science the map ultimately represents.

Part of the power of maps comes from their association with power, with knowledge, control, mastery, with the gaze (Jacob 2006, 318–20). And the suffragists were certainly wielding this power. But this also is about what Christian Jacob terms the "performative power" of maps, the idea that the experience of walking "through" a map confers "a feeling of symbolic mastery that often mirrors the power of kings and administrators" (Jacob 2006, 44). With "living suffrage maps" in parades and pageants, women became the map. In New York City, suffragists and non-suffragists alike could explore this performative power through the "Hopperie Game" at Luna Park, an amusement park were participants would: "hop on one leg up an incline over a map of non-suffrage states on to the suffrage states painted yellow, with the four ... states [voting on suffrage amendments in 1915] featured at the top and painted blue" (Finnegan 1999, 189; Women Open "Hopperie" 1915).[30] For a nickel, men, women, and children could contribute to the cause while moving their bodies across the states, embodying the spread of suffrage from the West to the East.

For suffragists, enacting the suffrage map and its evolution meant they embodied the changes they were working for. By becoming the map, they took it from a two-dimensional frozen moment-in-time to an interactive experience. By becoming the map in public, women could be empowered but they could also appeal to their audience. Viewers might see familiar faces – family members, neighbors, respected members of their community – in the roles of the states without the vote, in black and perhaps shackled. To take on the role of the state could definitely be a risky venture, however: hikers, parades, soapbox speakers were attacked, and the women not only berated but assaulted (Borda 2002; Schultz 2010; Stovall 2013).

Resisting the suffrage map

> … every 'white' State on the suffrage map is in the weird and woolly West…. Woman suffrage has been adopted only by the crude, raw, half-formed Commonwealths of the sagebrush and the windy plains, whence have come in endless procession foolish and fanatical politics and policies for a generation or two.
>
> Taylor 1913

There was resistance to suffrage on many grounds. Whether arguing against suffrage by suggesting it had ties to socialism or on the basis of women voting as "unnatural," many anti-suffrage arguments were made using the suffrage map and reflecting on its spatial patterns, comparing and contrasting them to other patterns, such as the quote above which ties suffrage to the "weird and woolly West" and essentially to the concept of frontier (Hubbard 1915).

A Mrs. William Forse Scott wrote to the *New York Times* in 1915:

> Feminism is rampant now and suffragists are boasting of alleged successes all along the line. It is time to face the danger. The solid block of suffrage States west of the great river, with one Eastern State added, stares at us from the suffrage map…. That suffrage States begin to measure their blunder is shown by open admission of the failure of the experiment.
>
> Scott 1915

Scott is clearly referencing the suffrage map with the slogan "The Map Proves It! Suffrage a Success…."

An anti-suffrage letter to the editor of the *New York Times*, commented on the similarities between the presence of the Mormon church and the passage of suffrage, suggesting readers consider the populations of Mormons in Western states and "compared with the suffrage map given in THE TIMES a few weeks ago" (Mormons as … 1913). The author was not only anti-suffrage but also anti-Mormon: by comparing the two, they were aligning suffrage with the Mormons and their more controversial beliefs such as polygamy, even though polygamy had been renounced by the main body of Mormons in 1890.

In 1912, an anti-suffrage organization reprinted a WCTU map, commenting in the caption:

> Many suffrage speakers insist that if the franchise is given to women they will be for prohibition. This map issued by the Women's Christian Temperance Union [?] shows plainly what we have stated before, that the six states having full suffrage – Colorado, Wyoming, Utah, Idaho, Washington, and California have not voted for prohibition. Not one of the states that has prohibition is a suffrage state.
>
> Prohibition Map ... 1912, 15

Just as the argument for suffrage could be made with the map, the argument against suffrage was made using the map. Its pervasiveness in the American landscape, the familiarity of the American public with the map, lent itself to these purposes.

"A cloudless map"

In 1913, Anna Howard Shaw, president of NAWSA, gave a speech at Carnegie Hall on the movement's progress entitled "A Cloudless Map" capturing the map in motion:

> The first map will be as black as Egypt. Then two little gray specks will show the beginning of school suffrage. The first glow of dawn will come in the West – first in Wyoming. The whole western part of the United States and the Pacific Coast will gradually be illuminated with the golden light of victory. Succeeding maps will show the gold extending in great floods of color after the elections of 1913, 1914, and 1915, until, at the beginning of 1920, only one black spot is left – namely, Vermont, where a constitutional convention is possible only once in ten years, and where the constitution cannot be amended except at a convention. The United States as the end of 1920 will have a cloudless map, all gold and no black, with a border of red, white and blue.
>
> Moving Map at Carnegie Hall 1913[31]

The moving map was a prediction that came true. In 1919, Congress approved the woman suffrage amendment, and by 1920 the amendment had been ratified by the required three-fourths of the states (Sims 1995, 333). Women had won the vote. Once suffrage was passed in the United States, the movement shifted. The suffrage organizations became the League of Woman Voters, an organization still existing today, and focused on educating women voters on issues. Women also worked internationally for the vote, through organizations such as the Women's International League for Peace and Freedom.

Through the suffrage map, suffragists were giving "tangible form to their beliefs" (Finnegan 1999, 8). Eventually, the persuasive image of the suffrage map was familiar enough to be invoked verbally, without having to be seen, and even remembered as suffragists looked back on their struggle: "I remember that there was a suffrage map reproduced from time to time" (Bompas 1942, 175). But it was not just a suffrage map, it was multiple suffrage maps, capturing the changing political terrain, as well as the inventive media on which the map appeared – posters, fans, drinking glasses, parade floats – and its placement across the American landscape, even into the halls of geographical societies. From its earliest version in 1907 through the ratification of the 19th Amendment in 1920, suffragists took the map to the streets, into people's homes, into horse races and ball games, into the halls of government. I cannot imagine a single American map produced, reproduced, or displayed as much as the suffrage map.

The suffrage map was consciously crafted to persuade viewers of the logic of the spread of suffrage. The color scheme was shifted to align the colors with the rhetoric, deliberately created to persuade. Suffragists were very much aware of the "ocular proof" maps provided and depended on their viewers to accept their maps as accurate, relying on the power of the simple choropleth map.[32] The familiar outline of the nation, the familiar cartographic conventions of line, shading, labels, key, are employed to counter the "appearance" of propaganda. As much as the suffrage movement included the map as part of their propaganda, they were relying on their audiences to perceive the suffrage map as a map and to accept its message. Consider for example Figures 4.4, 4.5, and 4.6. The suffrage map becomes yet another map created by Progressive women activists in their work. Its simplistic map structure echoes many of the maps created, published and distributed during this time period. It serves as a form of visual demonstration – conveying information best presented visually (Birdsell and Groarke 2007, 105).

Further, audiences were told to look at the map. A Nazi propaganda map reproduced in Hans Speier's (1942) article "Magic Geography" depicts the Allied troops surrounded by the Germans and was dropped from German airplanes to Allied troops in Belgium. Its text reads "Look at this map: it gives your true situation! Your troops are entirely surrounded – stop fighting! Put down your arms!" (Speier 1942, 328–9). Suffragists employed this tactic before the Nazis, telling their audiences to "… look upon the map… . Can you look upon the map unmoved?," "Our plea is simply that you look at the little suffrage map," and "'Look at New York State all in mourning'" (Harper 1922a, 429; One-Minute Talks … 1915; Ryan 1912).

We also must remember the suffrage map was always in the broader context of suffrage rhetoric, serving as a visual support ("ocular proof") to the written and oral rhetoric wielded in the campaign for the vote. It appeared extensively in suffrage publications (such as in *The Woman Citizen*, in the suffrage blue book, and in pamphlets) as well as being distributed as a flyer and as a poster. But it also was reprinted across the country in newspapers to capture the advance of suffrage, such as *The New York Times* in 1913, *The Tacoma Times* in 1914 (Washington State), *The Labor*

World and the *San Antonio Express* in 1916, *The Charlevoix County Herald* in 1918 (East Jordan, Michigan), and *The Review* in 1920 (High Point, North Carolina). While we might argue newspapers were using the suffrage map as a "non-propaganda" map, in fact, newspapers did run propaganda maps, such as the Mormon Octopus map published in the *Los Angeles Times* in 1904. The association of newspapers with "truth" and objectivity is a relatively new concept, really only taking hold in the early twentieth century (Soderlund 2002, 442–456). Maps like the suffrage map make concrete their subjects, as all thematic maps do: "… it's on a map, so it must be real" (Monmonier 1991, 88).

It was not just a single suffrage map, it was the consistent employment of the suffrage maps from 1907 to 1920 and the evolution of the image over time – the moving map, the ever-changing map. From the first map (Figure 5.1) to the "last suffrage map," the maps build on each other, visually capturing the success of the campaign from the first four suffrage states in the west to the completed map of 1920: "each clip builds on the previous one to form a coherent and often powerful composite impression" (Pickles 1992, 208). The newspaper headlines accompanying the maps capture this sense of movement: "How the Map of North America is Growing White" (*The Labor World* 1916).

"3436 blots of shame on the United States": Antilynching maps as propaganda maps

> The thousands of spires on churches of every denomination, running high into the heavens bear testimony that this is a Christian nation, or at least purports to be, and yet, actual records tell us that within the last thirty years we lynched 3,436 human beings – 3,436 blots of shame on the United States. Most of these lynchings occurred in the so-called, Solid South, bringing disgrace upon the entire Southern people, and condemnation from God and man.
>
> Harring 1922

Another political arena where American women wielded maps in order to attempt political transformation was in their effort to end the practice of lynching. The origins of the word "lynch" are contentious but date back before 1800 and are used to refer to the concept of "lynch law" or punishment without trial. As Michael Ayers Trotti pithily writes: "Stereotype of a lynching can be invoked with eight words and two commas: a mob, a nose, a swinging body defiled" (Trotti 2014, 852). But Trotti goes on to point out the spectrum of contexts and impacts are, in reality, far greater and much more complicated. Lynchings were used to punish men and women who crossed barriers, both physically and socially. Conducted by vigilante groups – groups taking the law into their hands, usually the dominant social group – lynching was used to keep subordinate groups in line. While it is often assumed lynching means hanging, lynching as a practice also includes shooting and burning (Miller 2005, 278). People were lynched for a wide variety of offenses with the most common being given as "rape" of a white woman. In reality, the real reason for the punishment might be "offenses" as shoplifting, disorderly conduct, "disrespecting" a white woman, contradicting a white man, unpopularity, or such "threatening" behaviors as registering to vote or running a successful business (Royster 1997, 28–9; Wells Barnett 1901; Zangrando 1980, 4 and 8). Lynchings were less common before the Civil War: it made more sense to sell slaves than kill them, if your slave did commit an offense you could be legally compensated for your loss, and mob violence undermined the plantation owners (Royster 1997, 10). After the Civil War and the abolition of slavery, lynching became a way for white southerners to keep former slaves and their sympathizers "in their place." It was less about actual crimes and more about a "cultural ideology" allowing "… them to define lynching and mob violence not as terrorism and race and gender control but as a 'right' action to avenge their honor, their manhood, their women" (Royster 1997, 32). But lynchings were not confined to the South, nor was this only a Black problem.

It is estimated between 1882 and 1968, 3,445 black men and women were lynched.[33] Smaller numbers of men and women of other ethnicities were also lynched, including white, Hispanic, and Asian. But by far, the most victims were Black American males and most were lynched in the South, although lynchings occurred across the country, including in my city of Omaha, Nebraska (Cook 2012, 55; Miller 2005, 277–8).

As the number of lynchings escalated, Black and White Americans became increasingly concerned and began to call for a stop to this racial terrorism. Towards ending this practice, the Tuskegee Institute, the NAACP, and the *Chicago Tribune* all compiled data and statistics on lynching incidents in the late nineteenth and early twentieth century (Pfeifer 2014, 844). This was part of a nationwide statistical movement in the United States originating with the federal government and its census data and extending to Americans interested in economic conditions, such as the leaders of industry who were members of organizations such as the American Geographical Society (Morin 2011, 31–8). While businessmen were more interested in data that could be used to advance their business interests, other Americans were interested in "moral statistics" (pauperism, alcoholism, prostitution, mental illness, incarceration, levels of education) in order to advance their social agendas. Statistics on lynchings were a form of moral statistics. When in the 1930s, a Southern Commission on the Study of Lynching began to seriously consider the practice, they turned to data compiled by the Tuskegee Institute (Raper 1933, 42). Efforts of halt the practice of lynching focused on two fronts: on shifting social tolerance of the practice and getting Congressional support to end the practice.

One of the most prominent early American anti-lynching spokespersons was Ida B. Wells-Barnett. Wells was raised in the South, educated, and was

a school teacher before she began writing for newspapers, eventually becoming editor and part owner of a successful Black newspaper in Memphis.
When three Black Memphis men were lynched, their businesses being more
successful than a neighboring white business, she called for a mass exodus
of Black Americans from Memphis and some did leave for the American
West, impacting local businesses. In this atmosphere of heightened racial
tensions, Wells wrote an editorial about lynchings for rape, suggesting the
then radical notion some White women chose to be in consensual relationships with Black men. Her reportage created a massive upheaval, with a
White mob attacking and destroying her press while she was luckily out of
town. Wells never returned to Memphis and eventually moved to Chicago
(Frisken 2012, 246).

I have not yet found any map images in Wells' numerous publications. But
she was very much aware of the spatial distribution of lynchings. As part of
her publications such as 1895's *A Red Record*, Wells tallied the lynchings by
state. She observed in 1892, of the 160 lynchings of Black Americans, 156
had occurred in southern states and that this was a region under "lawful ju-
risdiction" while in the Far West where you might find "greater lawlessness,"
there were comparatively fewer lynchings (Royster 1997, 28; Wells Barnett
1895, 15).

In at least one instance, a map image was created in connection to her
efforts. In a political cartoon from the 1895 *Richmond Planet*, drawn by
its editor John Mitchell Jr., Wells' success at bringing British attention to
American inaction on southern lynchings is captured.[34] In the cartoon, the
cannon of British public opinion is firing across the ocean at the US. Other
European leaders as well as in Africa and other parts of the world look on as
President Cleveland is pointing to the "outrages in Armenia" while behind
him, Wells lectures to audiences in the North and in the South and "bodies
of colored men may be seen hanging to trees" (Frisken 2012).

After Wells married, she was less active in campaigning against lynching, but active in woman's suffrage, even slipping into the Washington D.C.
suffrage parade of 1913 to march with her Illinois compatriots, rather than
segregated at the end of the parade (Alexander 1995, 79).

A more concerted effort to end lynching was long carried out by the
National Association for the Advancement of Colored People (NAACP).
The NAACP was formed in 1909 as "an interracial group determined to
safeguard black rights and challenge mob violence" after some especially
vicious riots roared through Springfield IL (Zangrando 1980, vii and 22).
Founders included such prominent American activists as Ida B. Wells, Jane
Addams, W.E.B. Du Bois, Sophinisba Breckenridge, John Dewey, Florence
Kelley, Walter Sachs, and Mary Church Terrell. Their principal objective
was to secure for all Americans the rights guaranteed in the 13th, 14th, and
15th Constitutional amendments: the end of slavery, equal protection under
the law, and universal male suffrage. Of significant concern was equal protection under the law – with the practice of lynching, accused persons were

being executed without a fair trial, without legal defense, without a jury of their peers determining their guilt or innocence. NAACP organizers were determined to expose injustices previously not been reported or largely ignored (Zangrando 1980, 23).

W.E.B. Du Bois establish the NAACP magazine *The Crisis* in 1910 and served as its editor. *The Crisis* had a regular column entitled "The Burden" where it documented and updated its audience on lynchings, reporting incidents and keeping a running tally. Eventually, the NAACP employed maps to visual capture the spatial distributions of lynchings, using the data tallied in their monthly columns. The first NAACP maps I've located were published in the 1919 publication *Thirty Years of Lynching in the United States 1889–1918*. The maps are relatively simple choropleth maps, with simplistic hand-drawn shading on the map, ranging from an "x" pattern for 1–24 lynchings to solid black for over 100 lynchings. Their shading paralleled the shading used on the suffrage maps with black being associated with less desirable conditions, with the resulting maps looking remarkably like the suffrage map with the South being solid black. The maps are not discussed in the extremely brief text. The publication is composed of a brief opening essay "Summation of the facts" before providing a chronological account of one hundred lynchings, a statistical analysis of the lynchings, and then a "Chronological List of Persons Lynched in the United States" including location, race, and alleged crime. Like Wells, the NAACP wished to present the full evidence so readers could decide for themselves and see the overwhelming evidence that fewer than 20% of those of lynched had been accused of rape (let alone tried and convicted) and that the crime disproportionally affected Black Americans (Zangrando 1980, 41).

In 1922, another antilynching map was published in *The Crisis* with map credit given to Miss Madeline Allison, a member of *The Crisis* staff (Figure 5.6).[35] The caption at the bottom reads "Each dot on this map represents one of the 3,436 lynchings which took place in the United States between 1889 and 1921, a total of 32 years. The dots are all in the states where the lynchings occurred, but naturally they could not be placed in the exact locations of the events within the state boundaries." This dot map appears to have been created using a Rand McNally base map, with "Rand McNally" just visible in the lower right corner. Again, the map is not discussed but rather is offered as a visual to accompany the listing of lynchings in "The Burden" column.

Allison's map was reprinted and used by the NAACP in campaigning in support of the Congressional Dyer Anti-Lynching Bill. Introduced in 1918, the Dyer Bill defined "… a mob as three or more persons acting without authority at law and held them liable to prosecution in federal court for a capital crime" (Zangrando 1980, 43). Officials who allowed a lynching to occur or failed to prosecute lynchers would face imprisonment and fines. And members of lynch mobs would be barred from serving on juries trying any cases under the act. The Bill passed in the House of Representatives in 1922 but was blocked by a Southern Democratic filibuster in the Senate.

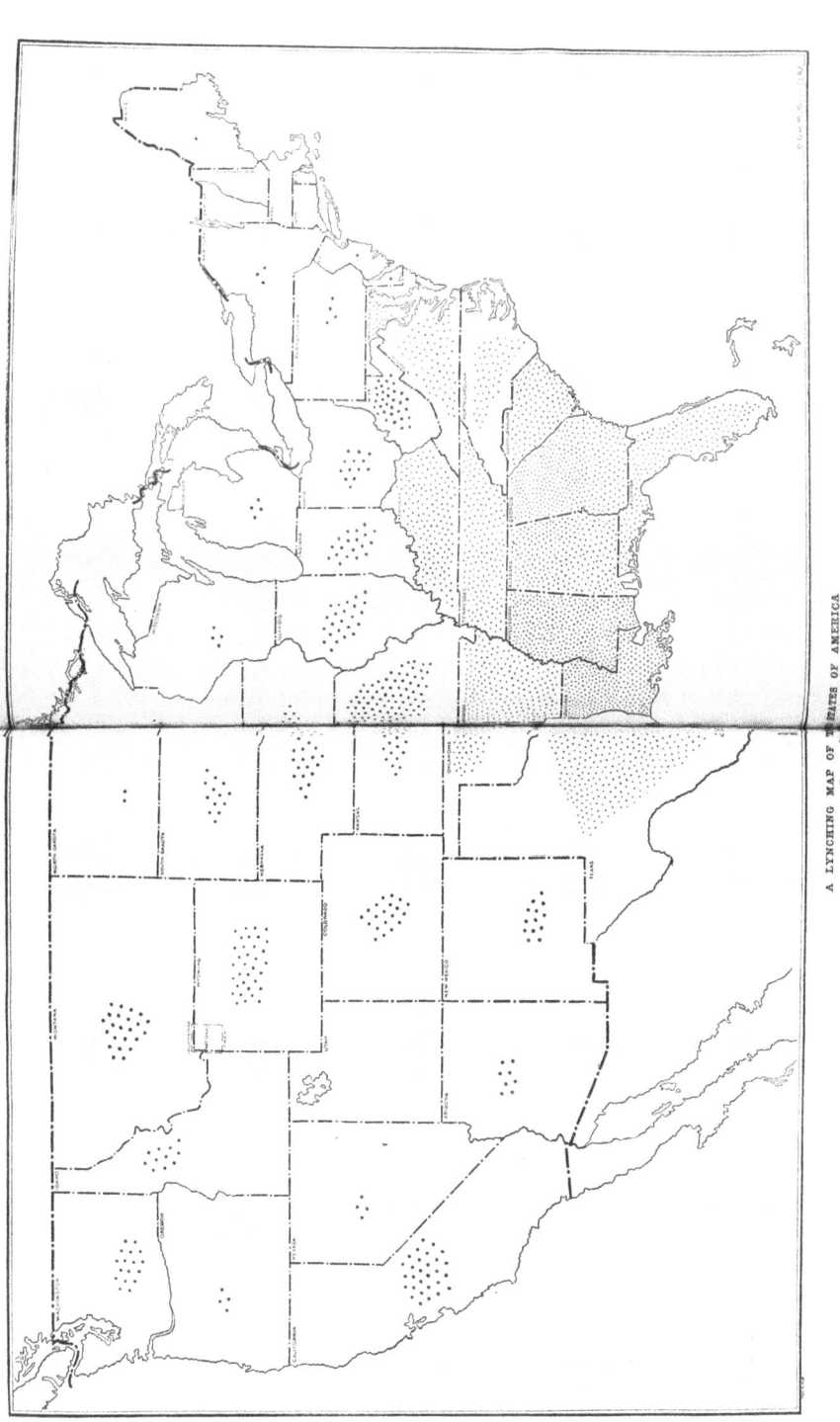

A LYNCHING MAP OF THE STATES OF AMERICA

Drawn by Miss Madeline Allison

Each dot on this map represents one of the 3,436 lynchings which took place in the United States between 1889 and 32 years. The dots are all in the states where the lynchings occurred, but naturally they could not be placed in the exact localities of the 32 states boundaries.

Figure 5.6 "A Lynching Map of the States of America. Drawn by Miss Madeline Allison" from the NAACP's *The Crisis* 23, 4 (1922): 168–169. Note the care with which the dots were drawn in the Southern states, apparently on a Rand McNally base map. The credit to Rand McNally is just visible in the lower right of the map. By permission of The Modernist Journals Project (searchable database). Brown and Tulsa Universities, ongoing. www.modjourn.org. Accessed 20 May 2017.

Subsequent bills were introduced to the U.S. Congress but never passed, always being blocked by Southern politicians. Congress never outlawed lynching.

Among the NAACP's papers, specifically those on their major campaigns, Allison's map and references to it appear, repackaged into a leaflet. The map now has a title: "3436 Blots of Shame on the United States: 1889–1922," with the phrase "blots of shame" suggesting to viewers that lynchings should be perceived as such. On the reverse of the leaflet, there is information about the Dyer Anti-Lynching Bill. It is likely this leaflet was distributed to Congress as they sought to gain allies. It also appears they distributed the map to newspapers for publication.[36] In the NAACP archives is a brief letter of support from Benjamin Ide Wheeler, President Emeritus of the University of California, with a handwritten added note: "Send 50 '3436 Blots of Shame.'"[37] The NAACP's maps appear to have been used to mobilize its members and readers to action through their reports and, placing their reports in national newspapers, were also trying to reach a broader audience on this major civil rights problem.[38]

In addition to the NAACP, there was a slightly later movement from a very different front, also attempting to end the practice of lynching. In 1930, a Texas suffragist named Jessie Daniel Ames began a campaign to organize southern white women against lynching. Moved by a case where a Texas mob dragged a black man from a jail cell and hung him, Ames sought to mobilize Southern women: "the wives and daughters of those who lynched" and those supposed being protected from the "sexual aggression" of black men (Hall 1977).[39]

Disturbed by an increased in lynchings, and by the lack of women on the Southern Commission on the Study of Lynching, Ames and other leaders of women's church and civic groups, such as the Council of Southern Methodist women, the Southern Baptist Missionary Union, the National Council for Jewish Women, the Y.W.C.A., the State Federation of Women's Clubs, began to organize (Working … 1933). The National Association of Southern Women for the Prevention of Lynching was never very large but claimed to represent the viewpoint of the educated, middle-class women of the South through "the 109 women's groups endorsing the anti-lynching campaign and on the 44,000 individuals who signed anti-lynching pledges" (Hall 1977). There was a conscious decision to limit the Association to white women (Miller 1978, 272). The association was not a "true" organization, not having a constitution, charter or even members but rather sought to work with existing organizations "to discuss the lynching problem and to develop ways to combat it" (Barber 1973, 380). Through existing organizations, they sought to have the organizations' members sign pledges denouncing lynching and stating they would work to enlighten public opinion (Barber 1973, 382). ASWPL focused on educating the public and preventing lynching, publishing and distributing pamphlets, leaflets, and fliers as part of their education program.

"This Business of Lynching"

Bulletin No. 4

Prevented Lynchings **55** •
Lynchings . . . . **14** +
(1934)

JESSIE DANIEL AMES
DIRECTOR

Association of Southern Women
for the Prevention of Lynching

January, 1935

Figure 5.7 Example of an Association of Southern Women for the Prevention of
Lynching bulletin with a map on its cover: "This Business of Lynching,"
1935, by permission of the Charles Deering McCormick Library of
Special Collections, Northwestern University Libraries (Political
Pamphlets P 201748).

Many of their brochures featured maps on the cover and, at times, inside
(Figure 5.7). The brochures produced by the Association were similar to the
publications of Ida B. Wells, and the NAACP: there are accounts of lynchings by state, there is some discussion of lynchings by time period, there are
statistics and graphics about the alleged crimes leading to lynchings. For
their maps, the ASWPL initially used available information, such as from

the *Negro Year Book* published at the Tuskegee Institute. But as the ASWPL grew, members began to make firsthand investigations of lynchings and attempted lynchings, as well as documenting prevented lynchings (Barber 1973, 383). As they investigated, they began to tabulate their own evidence and report it, establishing for themselves that while defending white women was the excuse, "the real cause lay in social and economic competition be-tween the two races" (Barber 1973, 384).

The maps depict the number of lynchings by state, with the lynchings depicted with a small cross, as if marking a grave. Where Ames and the As-sociation differ from Wells and the NAACP is on also tallying the *prevented* lynchings, which are also depicted on the maps with a dot (Slocum and Kessler 2015, 1514).[40] Education and public awareness were central to their approach but they also sought to empower Southern women to take action against lynchings, particularly by enlisting sheriffs to prevent lynchings. Through cases published in their brochures, such as in the brochure "Feel-ing Is Tense" (ASWPL 1938, 18), ASWPL publicized cases where lynchings were prevented, especially through the enlisting of law enforcement (Nor-dyke 1939, 685). Brochures from 1938 and 1942 switch from crosses and dots for lynchings and prevented lynchings to coloring states black where lynch-ings occurred. The total number of lynchings and prevented lynchings were declining, and the power of the crosses and dots were not as substantial as they had been even five years before. To create a visual with a stronger impact, they shifted to using just black and white: white for states without lynchings, black for states with lynchings.

The titles of the brochures function as slogans to the maps on their covers. Titles such as "This Business of Lynching," "Death by Parties Unknown," and "Feeling is Tense," serve to link the map to the social climate of the United States. Further, these titles are quotations: note the quotation marks around the titles, such as in Figure 5.7. These titles are repeated in the bro-chures, at times as headings, at times as part of narrative. For example, in the brochure "Feeling Is Tense," on the third page, the title is used as head-ing to a section describing the importance of county sheriffs and how they are very much aware of what the climate is in their counties: "If there is an intent to lynch a prisoner, the sheriff knows it and respond to the emotional demands of his supporters. He knows when 'feeling is tense' and judges ac-curately the number of votes involved" (ASWPL 1938, 3). This use of quota-tions as the titles to the brochures and to the maps changes slightly how the titles are read and interpreted. It shifts them from slogans to suggestions of what the public is thinking about the practice of lynching. Yet the critique implicit in the titles along with the maps elevates the maps to propaganda as they seek to sway their readers to question the practice of lynching in the United States and to, if necessary, take action, whether it is starting a local group (often through churches), reaching out to their community (through other churches, their local newspaper, to local civic groups), and enlisting local law enforcement.

The ASWPL was only active for about ten years. Whether it was due to their efforts or just an evolving American society: "As of May, 1940, the South, for the first time since records had been kept, went twelve months without a lynching." By 1941, the Association was dying. Some of its Executive Council believed their goal had been achieved: lynching had not been entirely stamped out but the number of lynchings was down and "in defense of women" was no longer an accepted justification (Barber 1973, 386–7). With the development of WWII, women's organization were focused on the war effort and the Association was dissolved in 1942.

The use of maps in the movement to end lynchings represents another effort by women to change the American landscape using maps, but which rode on the coattails of the suffrage movement. Jessie Daniel Ames acknowledge in 1942:

> ... woman suffrage made its contribution to the decrease in lynching and the changing of public opinion. For these women spoke as citizens who had the assurance born of the knowledge that they had the power to affect the political lives of local and county politicians whose bread and butter depended upon the will and the wishes of their constituents.
>
> Ames 1942, 60–1

The antilynching campaign never became as large as the suffrage movement and the map was never as integral to their rhetoric and iconography as the suffrage map. But ASWPL's use of maps was clearly in the realm of propaganda. Of the eight ASWPL pamphlets I have located, three have maps on their covers depicting lynchings and prevented lynchings. Two more have additional maps capturing the regional breakdowns of lynching by race and one has maps of lynching by decade. Unfortunately, the maps are left to speak for themselves: the maps are never discussed and the pamphlets' texts focus on statistics, specific accounts of lynchings, and accounts of prevented lynchings.

The NAACP and the ASWPL both used maps as part of their campaigns to end the practice of lynching. Both organizations had members with ties to the suffrage movement and were likely familiar with the pervasive American suffrage map and its use as part of suffrage propaganda. While neither groups' maps became as ubiquitous as the suffrage map, they were definitely borrowing a page from the suffrage campaign in attempting to provide ocular proof of the unequal rights and the need to "make the map white."

"The inexorable logic of geography": A map to change

Finish the Fight! We supply the Ammunition.
Esther Ogden, President of the National Woman
Suffrage Publishing Co., Inc.
Ogden 1919, 171

It is absolutely proven by the inexorable logic of geography, that equal suffrage works well. No one who looks at the map of the United States can deny that proof.

Votes for Women a Success ... 1913

Esther Ogden, in her "Report of the National Woman Suffrage Publishing Co., Inc." of 1919, documents the quantity of propaganda they produced "we printed over five million pieces of literature ... even with our small capital and limited facilities we have printed and distributed over fifty million pieces of literature in the five years since in the Company was incorporated" (Ogden 1919, 170–171). While the company was created to meet the needs of the growing suffrage movement in the United States, they also filled orders from England, Canada, South America, Mexico, Puerto Rico, and the Philippines. The suffrage map was part of the ammunition in their battle for the vote. The antilynching movement also used maps as part of their ammunition.

In the case of the suffrage map, it is possible to trace the evolution of the map from a rather straightforward depiction of states with and without suffrage to an instrument of suffrage propaganda. Over a period of years, the map was crafted to better convey the suffrage message, highlighted with slogans that emphasized the message, and then distributed across the country in a wide variety of formats. Eventually, the map was so widely known suffragists would refer to the map without it, referencing the pattern of white and black they assumed their audience was familiar with.

In contrast, the antilynching maps by the NAACP and the ASWPL walk a fine line between being factual and being propaganda. With Madeline Allison's map, the map itself is not altered to shift it towards propaganda, except to add the provocative title "3436 Blots of Shame." With the title, the dots indicating the number of lynchings shift from merely documenting events to being "blots of shame" on the map of the United States. With the ASWPL maps, the use of crosses, associated with graves, ties the marks on the map with actions on the ground, clearing linking the data with the loss of lives. In both cases, suffragists and antilynching activists were bent on making changes in the landscape, to transform their maps from black to white, from having "blots of shame" or "black spots" to being "pure" white.

Suffragists and antilynching activists through their employment of maps were essentially practicing critical cartography, using the maps associated with masculine surveillance to persuade male voters of the need for social reforms that would transform the American landscape. Maps have long been the communicators of "an imperial message ... used as an aggressive complement to the rhetoric of speeches, newspapers and written texts" (Harley 1988, 57). As women worked on social justice issues and sought to achieve political ends, they adopted political means to advance their cause, including maps, as well as devising their own methods, changing the practice of politics in the United States. They crafted and refined the image of the map to bring it in line with suffrage rhetoric. In this example of cartographic practice, we have "lay cartographers" using a fairly simple thematic map to make a compelling argument, then disseminating this image nationwide. After being subjected to mapping (mapped, as it were), women were employing simple mapping, on paper and with their own bodies, to advance their own agenda.

The radical feminist Audre Lorde famously wrote "the master's tools will never dismantle the master's house" (Lorde 1984, 123). But in this case, and in a clear case of critical cartography, the suffragists and anti-lynching activists are using the tool of cartography, wielding this instrument of power, to question social norms and argue for a greater role in American society as equal citizens and equal claimants of public space (Finnegan 1999, 49). In both cases, the map provided the "ocular proof" or "inexorable logic of geography" that there was a deep discrepancy in the rights Americans possessed between some parts of the country and others. The maps, whether of suffrage or of lynchings, served to make visible the unequal rights of Americans while arguing to rectify these inequalities.

Notes

1 Emphasis is Tyner's.
2 Note that American women in favor of suffrage were called "suffragists," not suffragettes. This was a tactical decision to distance the American suffrage movement from the more militant British suffrage movement (Sewell 2003, 97). Two works on the British suffrage movement that I have found particularly useful and admire greatly are: M. DiCenzo, L. Delap and L. Ryan, *Feminist Media History: Suffrage, Periodicals and the Public Sphere* (London: Palgrave Macmillan, 2011) and L. Tickner, *The Spectacle of Women: Imagery of the Suffrage Campaign 1907–1914* (Chicago, IL: University of Chicago Press 1988). Unfortunately, there are no works like these addressing the American suffrage movement.
3 This quote is referencing a popular suffrage poster created by B.M. Boye that won a suffrage poster contest and was thus used extensively in California's 1911 campaign. The striking poster features a young suffragist in front of the Golden Gate bridge and the setting sun, holding a Votes for Women banner, the sun creating a halo around her head (Sewell 2003, 85).
4 Finnegan discusses a similar campaign in downtown Minneapolis in 1917 (Finnegan 1999, 45).
5 *The Woman's Journal* 39, 3 (January 18, 1908): 12.
6 *The Woman's Journal* 39, 5 (February 1, 1908): 19. Anna Nicholes, in addition to being active with the suffrage movement, was involved with settlement houses in Chicago, with the Women's Trade Union League and with the Women's City Club.
7 Knobe was a journalist who had been on staff of the Chicago Tribune in the mid-1890s and who's biography in the *Woman's Who's Who of America* describes her as a "professional protégé of William T. Stead of London" (Leonard 1914–1915, 464). Stead used a map in his investigation of vice in Chicago, published in 1894. While conducting his investigation, Stead stayed at Hull-House.
8 Efforts to locate the map among the Rand McNally collections at The Newberry Library have not been fruitful. It seems Rand McNally moved offices three times over the course of the twentieth century and it is likely files were purged,

especially of contract work. Personal correspondence with Patrick Morris, The Newberry Library, 26 July 2016.

9 *The Woman Voter* 7, 4 (April 1916): 15.

10 Ibid.

11 *The Woman's Journal* clearly states in early 1908 they had a pamphlet available on the school vote featuring the suffrage map. Among the National Woman's Party Papers is a copy dated 1908 (available as a scan of the microfilm on ProQuest History Vault). However, a more widely available digital copy, made from a microfilm collection entitled "History of Women," dates their copy as 1913. It is the same pamphlet. The map is identical to map published by Knobe and republished in the 11 January 1908 issue of *The Woman's Journal*. By 1913, six states had passed woman's suffrage and the suffragists were adamant about making sure their maps were up to date (if not the minute!). I believe the correct publication date for the microfilm/digital pamphlet is 1908.

12 This map is reproduced in M. Wheeler's *One woman, one vote* (1995).

13 The first map located that has suffrage states in white with "man suffrage" in black is from *The Woman's Journal* 42, 30 (5 August 1911). It is not a propaganda map as such but is juxtaposed with a map depicting the legislative success for more popular government, termed "The Initiative and Referendum." An accompanying essay written by George Judson King of the National Referendum League discussed state by state their battles, not unlike the suffrage's movement's state-by-state documentation of success and failure. A similar version appears in Bertha Rembaugh's *The Political Status of Women in the United States* (1911).

14 In public maps created during Mussolini's reign, "the proportion of white to black increases to reflect the expansion of the empire…. The colours also served to remind viewers of the Fascist government's claim that it was bringing culture and civilization to even the darkest and most barren reaches of the empire, a common theme of the reign's colonial rhetoric" (Minor 1999, 149).

15 Flyers produced for the California campaign were translated into German, French, and Italian to reach all of California's diverse population (DuBois and Kearns 1995, 16). To date, I have only found the suffrage map in English.

16 For more information on suffrage and consumer culture, particularly their wielding of advertising, see Margaret Finnegan's excellent study *Selling Suffrage: Consumer Culture & Votes for Women* (1999).

17 Calendars mentioned in "Big suffrage party for Mrs. Whitman" *New York Times* 10 October 1915; fans mentioned as part of the New York campaign (specifically "35,000 fans carrying the suffrage map") in Harper 1922, pp. 471; a fan with a suffrage map reading "While watching the Horse Race, Don't forget one half the Human Race," Massachusetts campaign 1915, can be seen on the following website: https://lewissuffragecollection.omeka.net/items/show/989, accessed 20 May 2017; drinking glasses are described in "Wives of Soldiers" 1916; and maps in baseball programs are described in "Suffrage Leaders" 1915.

18 Today, occasionally copies of the hand fans with the suffrage map printed on them turn up on online auctions such as on ebay.

19 The American Geographical Society Library has two poster-size suffrage maps in their collections.

20 The Oklahoma map is reproduced in R. Cooney's *Winning the Vote* (2005).

21 The Pennsylvania Men's League for Women Suffrage can be seen online at: http://digitallibrary.hsp.org/index.php/Detail/Object/Show/object_id/9952. Accessed 30 December 2016.

22 Empire State Campaign Committee district maps, Women's Studies Manuscript Collections from the Arthur and Elizabeth Schlesinger Library, Radcliffe College, Series 1: Woman's Suffrage, Part B: New York (Helen Barten Owens A-147, item # 73).

23 This map is illustrated in Dando 2010, Figure 5.6.

24 Alice Paul went to England in 1907 and became involved in the Women's Social and Political Union, attending meetings, marching in parades, eventually arrested with other suffragettes, imprisoned, went on a hunger strike and was force fed. When Paul returned to the US in 1910, she was a minor celebrity (Stovall 2013, 12–15).

25 See also Walter 1913, which also discusses the parade plans and specifically mentions the float with "the suffrage map with 'the nine states of light among thirty-nine of darkness.'"

26 A photograph of a map float appears in Carrie Chapman Catt Albums at Bryn Mawr College Library Special Collections and was published in Dando (2010), Figure 5.4. According to the information on the Special Collections' website, the photograph was taken in Omaha in 1919. It appears to match the description of the Washington float. It is unclear if they were still using an un-updated float or whether the information in the album may be incorrect. The album caption states "Celebrating the 'Ratification' by the Ninth State," which would be Arizona in 1912, suggesting the date may reflect the album and not the photograph which likely was taken in 1912.

27 It is unclear whether Dock wore her map in the procession in Washington. For an image of Dock with her map and the other hikers, see: https://www.loc.gov/item/ggb2005012658/. Accessed 20 May 2017.

28 The albums/scrapbooks are held at Bryn Mawr College Library Special collections. The images is available for viewing at: http://triptych.brynmawr.edu/cdm/landingpage/collection/suffragists. Accessed 20 May 2017. One of the images captures Ms. Catt with tableau participants.

29 Emphasis is Schultz's.

30 The Hopperie game reportedly received extensive news coverage: thirteen city newspapers featured it in twenty-nine stories (Finnegan 1999, 189). Unfortunately, the photographs located to date do not depict the map. See, for example, a photograph on the Library of Congress' website: http://www.loc.gov/pictures/item/2005688140/. Accessed 20 May 2017.

31 Shaw's speech on "A Cloudless Map in 1920" was also covered in the *Plainfield Daily Press* (NJ) 12 February 1913: 7. While the text of many of Shaw's speeches are available in archives, this speech has not yet been located.

32 While choropleth maps can be misleading and can easily be manipulated by using inappropriate category breaks, the suffrage maps are not misleading in the sense they only depict states that have passed suffrage and those that have not (Tyner 2015, 1090–1). See also Crampton (2004).

33 At best, all we have are estimates of the number of lynchings that occurred. There are calls to create a "National Lynching Database," such as by Lisa Cook in her scholarly article "Converging to a National Lynching Datatbase" (2012). In February, the Equal Justice Initiative announced they had compiled statistics on "terror lynchings" in the South and had documented 3,959 instances. Their report received national news coverage.

34 The editorial cartoon, "Firing at Long Range" can be seen at: http://songswithoutwords.org/items/show/222. Accessed 31 December 2016.

35 A photograph of *The Crisis* staff from 1912 shows Miss Madeline Allison (*The Crisis* vol. 5, no. 1, page 35).

36 A version of the map that supposedly appeared in newspapers is on the National Endowment for the Humanities website: http://edsitement.neh.gov/curriculum-unit/naacps-anti-lynching-campaigns-quest-social-justice-interwar-years. Accessed 20 May 2017. Unfortunately, the creators of the website do not identify the newspapers and today's web-managers do not know where the map came from (correspondence with current website staff).

37 Note dated 15 February 1922. Papers of the NAACP, Part 07: The Anti-Lynching Campaign, 1912–1955, Series B: Anti-Lynching Legislative and Publicity Files, 1916–1955. http://congressional.proquest.com/histvault?q=001529-012-0861. Accessed 25 July 2016.
38 Allison's map was also used in a brochure published by the American Civil Liberties Union entitled "Mob Violence in the United States" (1923).
39 Ames is not the only suffragist to become involved in the antilynching movement. Elisabeth Freeman, an American active in the British suffrage movement before becoming involved in the American movement, was working on suffrage in Texas when Jesse Washington was lynched in Waco in 1916. The NAACP hired her to investigate, as part of their efforts to ascertain the exact events that led to lynchings. Freeman gathered information and photographs on the Waco events and sent them to the NAACP. DuBois took the information and crafted a supplement to *The Crisis* entitled "The Waco Horror." DuBois used the incident to kick off an antilynching campaign, hiring Freeman to travel the country giving presentations (Bernstein 2005).
40 The NASWPL maps appear to have been copied for a publication by the Southern Commission on the Study of Lynching. The map has been redrawn but depicts both lynchings and prevented lynchings as the NASWPL maps does (Raper 1933, 42). The author credits Tuskegee but not the NASWPL.

Primary Sources

Addams, J. 1910. Why women should vote. *The Ladies Home Journal* 27 (January): 21–22.
American Civil Liberties Union. 1923. *Mob Violence in the Unites States: The Striking Facts in Brief Presented by the American Civil Liberties Union.* New York: American Civil Liberties Union.
Ames, J. 1942. *The Changing Character of Lynching: Review of Lynching, 1931–1941.* Atlanta, GA: Commission on Interracial Cooperation, Inc.
Association of Southern Women for the Prevention of Lynching. 1935. *"This Business of Lynching".* Bulletin No. 4. Atlanta, GA: Association of Southern Women for the Prevention of Lynching.
———. 1936. *"Death By Parties Unknown".* Bulletin No. 6. Atlanta, GA: Association of Southern Women for the Prevention of Lynching.
———. 1938. *"Feeling Is Tense".* Bulletin No. 8. Atlanta, GA: Association of Southern Women for the Prevention of Lynching.
Big suffrage party for Mrs. Whitman. *The New York Times* 10 October 1915, p. 27.
The Black Belt. 1914. *Chicago Daily Tribune* 20 March, p. 6.
Blackwell, A. 1908. *Women and the School Vote.* New York: National American Woman Suffrage Association.
———. 1913a. The pageant procession. *The Woman's Journal 44, 9 (*1 March): 68.
———. 1913b. *Women and the school vote.* New York: National American Woman Suffrage Association.
Boggs, S. 1947. Cartohypnosis. *The Scientific Monthly* 67, 1: 469–476.
Bompas, K. 1942. Our background: struggle and achievement. *International Women's News* 36, 10: 175.
College Equal Suffrage League of Northern California. 1913. *Winning Equal Suffrage in California: Reports of Committees of the College Equal Suffrage League of Northern California in the Campaign of 1911.* San Francisco CA: National College Equal Suffrage League.

Dock, L. 1913. Walks and wins with two-ft. map. *The Woman's Journal* 44, 5 (1 February): 40

Equal Rites. Another interesting debate by the female suffrage agitators. *The New York Times* 14 May 1869.

Georgia vs. Connecticut. 1919. *The Woman Citizen: The Woman's National Political Weekly* 10 May: 1.

Harper, I. ed. 1922a. *History of Woman Suffrage*, Vol. 5: 1900–1920. New York: National American Woman Suffrage Association.

———. 1922b. *History of Woman Suffrage*, Vol. 6: 1900–1920. New York: National American Woman Suffrage Association.

Harring, J. 1922. A lynching! *Missionary Seer* 22, 12: 7–8.

How map of North America is growing white. 1916. *The Labor World* 4 November, 4.

Hubbard, B. 1915. *Socialism, Feminism, and Suffragism, the Terrible Triplets.* Chicago: American Publishing Company.

The jubilant diary of a Golden Jubilee. 1919. *The Woman Citizen* 3, 44: 924.

Knobe, B. 1907. The suffragists' uprising. *Appleton's Magazine* 10: 772–779.

———. 1908a. That suffrage map. *The Woman Citizen* 39, 11 (14 March): 42.

———. 1908b. Votes for women: an object-lesson. *Harper's Weekly* 52, 2679: 20-21.

Leonard, J. 1914–1915. *Woman's Who's Who of America: A Biographical Diction-ary of Contemporary Women of the United States and Canada.* New York: The American Commonwealth Company.

Logan, A. 1912. Colored women as voters. *The Crisis* 4, 5: 242–243.

Make Michigan 100% white on the suffrage map. 1918. *The Charlevoix County Herald* (East Jordan, Mich.) 6 September.

McMahon, A. 1918. How to win a state. *The Woman Citizen* 3, 25: 508–509.

Mormons as suffrage pioneers. *The New York Times* 29 May 1913.

Moving map at Carnegie Hall. 1913. *The Woman's Journal and Suffrage News* 44, 6 (8 February): 48.

National American Woman Suffrage Association. 1907. *Proceedings of the Thir-ty-Ninth Annual Convention of the National-American Woman Suffrage Associa-tion.* Warren, OH: The Association.

———. 1912. *Proceedings of the Forty-Fourth Annual Convention of the Nation-al-American Woman Suffrage Association.* Washington, D.C.: The Association.

———. 1914. *The Hand Book of the National American Woman Suffrage Associa-tion and Proceedings of the Forty-Sixth Annual Convention.* New York: National Woman Suffrage Publishing Company, Inc.

———. 1919. *The Hand Book of the National American Woman Suffrage Associa-tion and Proceedings of the Jubilee Convention (1869–1919).* New York: National American Woman Suffrage Publishing Company, Inc.

New leaflets. 1908. *The Woman's Journal* 39, 6 (February 8): 24.

Nordyke, L. 1939. Ladies and lynching. *Survey Graphic* 28: 683–686.

Ogden, E. 1919. Report of the National Woman Suffrage Publishing Co., Inc. In *Handbook of the National American Woman Suffrage Association and Proceed-ings of the Jubilee Convention*, ed. J. Wilson, pp. 169–171. New York: National American Woman Suffrage Association.

One-minute talks at suffrage shop. *The New York Times* 31 January 1915.

Parade to glow like rainbow. 1913. *The Woman's Journal* 44, 8 (February 22): 57 and 62.

Park, A. 1908. Suffrage map enlarged. *The Woman's Journal* 39, 29: 113.

Prohibition map of the United States. 1912. *The Woman's Protest Against Woman Suffrage* 1, 6: 15.

Raper, A. 1933. *The Tragedy of Lynching.* Chapel Hill, NC: The University of North Carolina Press.

Rembaugh, B. 1911. *The Political Status of Women in the United States: A Digest of the Laws Concerning Women in the Various States and Territories.* New York and London: G. P. Putnam's Sons.

Residents of Hull House. 1895. *Hull-House Maps and Papers, a presentation of nationalities and wages in a congested district of Chicago, together with comments and essays on problems growing out of the social conditions.* New York and Boston: T.Y. Crowell.

Ryan, A. 1912. Map of America changes. *The Woman's Journal* 43, 12: 91.

Scott, W. 1915. Suffrage a failure. *New York Times* 24 January 1915.

Speier, H. 1942. Magic geography. *Social Research* 8: 310–330.

Stapler, M., ed. 1917. *The Woman Suffrage Year Book 1917.* New York: National Woman Suffrage Publishing Company, Inc.

Stead, W. 1894. *If Christ Came to Chicago! A Plea for the Union of All Who Love in the Service of All Who Suffer.* Chicago, IL: Laird & Lee, Publishers.

Suffrage leaders get their innings. *The New York Times* 19 May 1915.

A suffrage map. 1908. *The Woman's Journal* 39, 2 (January 11): 5.

The suffrage map today. 1914. *The Tacoma Times* 17 November, 5.

'Suffragists' machine perfected in all states under Mrs. Catt's rule. *New York Times* 29 April 1917.

Suffragists silent in park gathering. 1916. *The New York Times* 24 September 1916.

Taylor, W. 1913. No Eastern suffrage states. *The New York Times* 7 May 1913.

That suffrage map. 1908. *The Woman's Journal* 39, 5 (1 February): 19.

Tinnin, G. 1913. Why the pageant? *The Woman's Journal* 44, 7 (February 15): 50.

Vernon, M. 1914. Campaigning through Nevada. *The Suffragist* 2, 21: 6.

Virginia women want ratification. 1919. *The Woman Citizen* 4, 11: 266.

Votes for Women a Success: Proven By the Map of the United States. 1913. New York: National American Woman Suffrage Association.

Votes for women: they're a success – the map proves it. 1916. *San Antonio Express* 51, 107.

Wells Barnett, I. 1895. *A Red Record.* Chicago, IL: Ida B. Wells.

———. 1901. Lynching and the excuse for it. *The Independent* 53, 2737: 1133–1136.

Wives of soldiers seek financial aid. *The New York Times* 1 July 1916.

Woman suffrage. 1920. In *The Encyclopedia Americana: a library of universal knowledge*, pp. 446–454. New York: The Encyclopedia Americana.

The woman suffrage map of 1914. *The Madrid Herald* 24 December 1914: 2.

Women open "hopperie." 1915. *The New York Times* 27 June 1915.

Woods, R. and A. Kennedy, eds. 1911. *Handbook of Settlements.* New York: Charities Publication Committee, The Russell Sage Foundation.

Working for a lynchless year. 1933. *The Literary Digest* January 7, p. 19.

Bibliography

Alexander, A. 1995. Adella Hunt Logan, the Tuskegee Woman's Club, and African Americans in the suffrage movement. In *Votes for Women! The Woman Suffrage Movement in Tennessee, the South, and the Nation*, ed. M. Wheeler, pp. 71–104. Knoxville, TN: The University of Tennessee Press.

Barber, H. 1973. The Association of Southern Women for the Prevention of Lynching, 1930–1942. *Phylon* 34, 4: 378–389.

Bar-Gal, Y. 2003. The blue box and JNF propaganda maps, 1930–1947. *Israel Studies* 8, 1: 1–19.

Bernstein, P. 2005. *The First Waco Horror: The Lynching of Jesse Washington and the Rise of the NAACP.* College Station, TX: Texas A & M University Press.

Birdsell, D. and L. Groarke. 2007. Outlines of a theory of visual argument. *Argumentation & Advocacy* 43: 103–113.

Blair, K. 1994. *The Torchbearers: Women and Their Amateur Arts Associations in America, 1890–1930.* Bloomington, IN: Indiana University Press.

Borda, J. 2002. The woman suffrage parades of 1910–1913: possibilities and limitations of an early feminist rhetorical strategy. *Western Journal of Communication* 66, 1: 25–52.

Boria, E. 2008. Geopolitical maps: a sketch history of a neglected trend in cartography. *Geopolitics* 13: 278–308.

Brantlinger, P. 1985. *Victorians and Africans: The Genealogy of the Myth of the Dark Continent.* Chicago, IL: University of Chicago Press.

Buechler, S. 1986. *The Transformation of the Woman Suffrage Movement: The Case of Illinois, 1850–1920.* New Brunswick, NJ: Rutgers University Press.

Cairo, H. 2006. 'Portugal is not a small country': maps and propaganda in the Salazar regime. *Geopolitics* 11: 367–395.

Chapman, M. 1999. Women and masquerade in the 1913 suffrage demonstration in Washington. *Amerikastudien/American Studies* 44, 3: 343–355.

Chávez, K. 2009. Remapping latinidad: a performance cartography of latina/o identity in rural Nebraska. *Text and Performance Quarterly* 29, 2: 165–182.

Churchill, R. 2002. *Mapping Chicago – Making Chicago* (Newberry Slide Set No. 34). Chicago, IL: The Herman Dunlap Smith Center for the History of Cartography, The Newberry Library.

Churchill, R. and S. Slarsky. 2004. Mapping September 11, 2001: cartographic narrative in the print media. *Cartographic Perspectives* 47: 13–27.

Cook, L. 2012. Converging on a national lynching database: recent developments and the way forward. *Historical Methods* 45, 2: 55–63.

Cooney, R. 2005. *Winning the Vote: The Triumph of the American Woman Suffrage Movement.* Santa Cruz, CA: American Graphic Press.

Crampton, J. 2004. GIS and geographic governance: reconstructing the choropleth map. *Cartographica* 39, 1: 41–53.

Dalal, F. 2002. *Race, color and the processes of racialization: new perspectives from group analysis, psychoanalysis and sociology.* New York: Brunner-Routledge.

Dando, C. 2010. 'The map proves it!': map use by the American woman suffrage movement. *Cartographica* 45, 4: 221–240.

Downey, D. 1987. William Stead and Chicago: a Victorian Jeremiah in the Windy City. *Mid-America: An Historical Review* 68: 153–66.

DuBois, E. 1995. Working women, class relations, and suffrage militance: Harriot Stanton Blatch and the New York woman suffrage movement, 1894–1909. In *One Woman, One Vote: Rediscovering the Woman Suffrage Movement,* ed. M. S. Wheeler, pp. 221–244. Troutdale, OR: NewSage Press.

DuBois, E. and K. Kearns. 1995. *Votes for Women: A 75th Anniversary Album.* San Marino, CA: Huntington Library.

Dumenil, L. 2007. The new woman and the politics of the 1920s. *OAH Magazine of History* 21, 3: 22–26.

Finnegan, M. 1999. *Selling Suffrage: Consumer Culture & Votes for Women*. New York: Columbia University Press.

Friendly, M. and G. Palsky. 2007. Visualizing nature and society. In *Maps: Finding Our Place in the World*, eds. J. Akerman and R. Karrow, pp. 207–253. Chicago, IL: University of Chicago Press.

Frisken, A. 2012. 'A song without words': anti-lynching imagery in the African American press, 1889–1898. *The Journal of African American History* 97, 3: 240–269.

Frost-Knappman, E. and K. Cullen-DuPont. 2005. *Eyewitness History: Women's Suffrage in America*. New York: Facts on File.

Hall, J. 1977. Women and lynching. *Southern Exposure* 4, 4: 53–54.

Harley, J. 1988. Maps, knowledge, power. In *The Iconography of Landscape: Essays on the Symbolic Representation, Design, and Use of Past Environments*, eds. D. Cosgrove and S. Daniels, pp. 277–312. Cambridge: Cambridge University Press.

Herb, G. 1997. *Under the Map of Germany: Nationalism and Propaganda 1918–1945*. London: Routledge.

Jacob, C. 2006. *The Sovereign Map: Theoretical Approaches to Cartography throughout History*, trans. T. Conley, ed. E. Dahl. Chicago, IL: University of Chicago Press.

Jarosz, L. 1992. Constructing the dark continent: metaphor as geographic representation of Africa. *Geografiska Annaler* (ser. B) 74: 105–115.

Jenkins, J. 2011. Marching shoulder to shoulder: new life in the Connecticut woman suffrage movement. *Connecticut History* 50, 2: 131–145.

Kimball, M. 2006. London through rose-colored graphics: visual rhetoric and information graphic design in Charles Booth's maps of London poverty. *Journal of Technical Writing and Communication* 36: 353–381.

Latour, B. 1987. *Science in Action: How to Follow Scientists and Engineers through Society*. Cambridge, MA: Harvard University Press.

Le Corbeiller, C. 1961. Miss America and her sisters: personifications of the four parts of the world. *Metropolitan Museum of Art Bulletin* (n.s.) 19: 209–223.

Lorde, A. 1984. *Sister Outsider: Essays and Speeches*. Berkeley, CA: Ten Speed Press.

Madsen, A. 2014. Columbia and her foot soldiers: civic art and the demand for change at the 1913 suffrage pageant-procession. *Winterthur Portfolio* 48, 4: 283–310.

McCammon, H., and K. Campbell. 2001. Winning the vote in the west: the political successes of the women's suffrage movements, 1866–1919. *Gender & Society* 15, 1: 55–82.

McCammon, H., K. Campbell, E. Granberg and C. Mowery. 2001. How movements win: gendered opportunity structures and U.S. women's suffrage movements, 1866 to 1919. *American Sociological Review* 66: 49–70.

McCammon, L. Hewitt, and S. Smith. 2004. 'No weapon save argument': strategic frame amplification in the U.S. woman suffrage movements. *The Sociological Quarterly* 45, 3: 529–556.

McGerr, M. 1990. Political style and women's power, 1830–1930. *The Journal of American History* 77, 3: 864–885.

Miller, K. 1978. The ladies and the lynchers: a look at the Association of Southern Women for the Prevention of Lynching. *Southern Studies* 17: 261–280.

Miller, R. 2005. Lynching in America: some context and a few comments. *Pennsylvania History: A Journal of Mid-Atlantic Studies* 72, 3: 275–291.

Minor, H. 1999. Mapping Mussolini: ritual and cartography in public art during the second Roman Empire. *Imago Mundi* 51: 147–162.

Monmonier, M. 1991. *How to Lie with Maps*. Chicago, IL: University of Chicago Press.

Moore, S.J. 1997. Making a spectacle of suffrage: the national woman suffrage pageant, 1913. *Journal of American Culture* 20, 1: 89–103.

Morin, K. 2011. *Civic Discipline: Geography in America, 1860–1890*. Farnham UK: Ashgate Publishing Ltd.

"The Movement Comes of Age." 2005. Texas State Library and Archives Commission. Available at: www.tsl.texas.gov/exhibits/suffrage/comesofage/page1.html. Accessed 19 June 2017.

Palczewski, C. 2005. The male Madonna and the feminine Uncle Sam: visual argument, icons, and ideographs in 1909 anti-woman suffrage postcards. *Quarterly Journal of Speech* 91, 4: 365–394.

Pavlovskaya, M. and K. St. Martin. 2007. Feminism and geographic information systems: from a missing object to a mapping subject. *Geography Compass* 1: 583–606.

Perkins, C. 2009. Performative and embodied mapping. In *International Encyclopedia of Human Geography*, eds. C. Perkins, R. Kitchin, & N. Thrift, pp. 126–132. London: Elsevier.

Petto, C. 2009. Playing the feminine card: women of the early modern map trade. *Cartographica* 44, 2: 67–81.

Pfeifer, M. 2014. At the hands of parties unknown? The state of the field of lynching scholarship. *The Journal of American History* 101, 3: 832–846.

Pickles, J. 1992. Texts, hermeneutics and propaganda maps. In *Writing Worlds: Discourse, Text, and Metaphor in the Representation of Landscape*, eds. T. Barnes and J. Duncan, pp. 193–230. London and New York: Routledge.

Reynolds, L.J. 2005. 'Strangely ajar with the human race': Hawthorne, slavery, and the question of moral responsibility. In *Hawthorne and the Real: Bicentennial Essays*, ed. M. Bell, pp. 40–69. Columbus, OH: Ohio State University Press.

Robinson, A. 1982. *Early Thematic Mapping in the History of Cartography*. Chicago, IL: University of Chicago Press.

Royster, J. 1997. Introduction. In *Southern Horrors and Other Writings: The Anti-Lynching Campaign of Ida B. Wells, 1892–1900*, ed. J. Royster, pp. 1–41. Boston/New York: Bedford Books.

Schoolman, M. 2014. *Abolitionist Geographies*. Minneapolis, MN: University of Minnesota Press.

Schulten, S. 2012. *Mapping the Nation: History and Cartography in Nineteenth Century America*. Chicago, IL: University of Chicago Press.

Schultz, J. 2010. The physical is political: woman's suffrage, pilgrim hikes and the public sphere. *The International Journal of the History of Sport* 27, 7: 1133–1153.

Sewell, J. 2003. Sidewalks and store windows as political landscapes. *Perspectives in Vernacular Architecture* 9 (Constructing Image, Identity, and Place): 85–98.

Sims, A. 1995. Armageddon in Tennessee: the final battle over the Nineteenth Amendment. In *One Woman, One Vote: Rediscovering the Woman Suffrage Movement*, ed. M. S. Wheeler, pp. 333–352. Troutdale, OR: NewSage Press.

Slocum, T. and F. Kessler. 2015. Thematic mapping. In *History of Cartography Volume 6*, ed. M. Monmonier, pp. 1500–1524. Chicago, IL: University of Chicago Press.

Soderlund, G. 2002. Covering urban vice: *The New York Times*, 'white slavery,' and the construction of journalistic knowledge. *Critical Studies in Media Communication* 19, 4: 438–460.

Stovall, J. 2013. *Seeing Suffrage: The Washington Suffrage Parade of 1913, its Pictures, and its Effect on the American Political Landscape.* Knoxville, TN: The University of Tennessee Press.

Terborg-Penn, R. 1998. *African American Women in the Struggle for the Vote, 1850–1920.* Cambridge, MA: Harvard University Press.

Tickner, L. 1988. *The Spectacle of Women: Imagery of the Suffrage Campaign 1907–1914.* Chicago, IL: University of Chicago Press.

Trotti, M. 2014. The multiple states and fields of lynching scholarship. *The Journal of American History* 101, 3: 852–853.

Tyner, J. 2015. Persuasive cartography. In *History of Cartography Volume 6*, ed. M. Monmonier, pp. 1087–1095. Chicago, IL: University of Chicago Press.

Wheeler, M. 1995. A short history of the woman suffrage movement in America. In *One Woman, One Vote: Rediscovering the Woman Suffrage Movement*, ed. M. S. Wheeler, pp. 9–19. Troutdale, OR: NewSage Press.

Wood, M. 1999. *Blind Memory: Visual Representations of Slavery in England and America 1780–1865.* New York: Routledge.

Woodward, D. and G. Lewis. 1998. Introduction. In *History of Cartography Volume 2, Book 3: Cartography in the Traditional African, American, Arctic, Australian, and Pacific Societies*, eds. D. Woodward and G. Lewis, pp. 1–10. Chicago, IL: University of Chicago Press.

Zangrando, R. 1980. *The NAACP Crusade against Lynching, 1909–1950.* Philadelphia, PA: Temple University Press.

6 Maps in motion, women in motion

Where is woman's place, anyway? Today she is flying across the ocean, swimming the Channel, running for mayor, writing the world's important books and doing the world's big jobs.

Stay at home, lady? No more! Travel … see … do … live. That's the program of modernity.

White Star Lines advertisement, *Good Housekeeping* 1928[1]

Then, of course, there are the maps—the map fliers, the victory maps, large fliers, the map posters of the U.S. and North America; but it's a wild race, my merrie suffragists, to keep up with the victories. Printer's ink is hardly dry before the whole output must go into the discard. Nevertheless, new maps and new literature you shall have, though your winded Literature Committee is heard to grumble under their breath "varium et mutabile semper *literature* est."

Livermore 1919, 174[2]

We speak of a "Geospatial Revolution" occurring in the twenty-first century, of a proliferation of maps in the public sphere thanks to geographic information systems, global positioning systems, cloud mapping, of a "people's cartography" in practice (Schnell and Leuenberger 2014, 518). But one hundred years ago, in the early twentieth century, a mapping revolution of sorts was in motion, one that was outside the halls of academia, government, and industry, involving everyday Americans using geography and cartography to argue for changes in the American landscape.

American women's lives were in transition: a "New Woman" had emerged and American women were taking on greater public roles on a variety of fronts. A rhetoric of "separate spheres" for women and men suggested that women and men occupied different spaces, but this was not the reality for most American women. Society might suggest woman's place was in the home, yet many women were quite active in public. In the Progressive Era, men but especially women took on a wide variety of social issues that they believed should be addressed but had not. As they worked on their social issues, they took on a greater public presence. American women's lives were in *motion.* Thanks to the bicycle and then the automobile, they were venturing farther from home and with greater autonomy. American women missionaries, whether domestic or international, traveled and worked and shared their experiences with audiences back in the United States. Social settlement workers lived in the communities they were trying to uplift, working on a wide-variety of levels with community residents, at times conducting door-to-door surveys, visiting ill residents, and other activities that took them out into the community. Suffrage activists traveled through their towns, their counties, sometimes throughout their states and even farther to reach audiences and advocate for the vote for women. Antilynching activists might give presentations, prevent lynchings by prompting intervention from law enforcement or be lobbying for legislative changes to end the practice of lynching. As American women became more mobile, they used and created geographic knowledge, at times creating maps to better communicate their vision.

The maps created by American women as part of their social activism were "in *motion*" too.[3] They shared information about how and where to cycle and drive, creating maps for these purposes and circulating these maps. American women missionaries worked to educate their at-home supporters on the geography of their mission countries, providing basic reference maps and urging over and over again the use of maps. Social settlement workers conducted studies, wrote reports and drafted maps, then circulated their reports and maps widely. Additionally, they promoted "social studies" and mapping as practices among other like-minded women; that in order to better assist their communities, they had to gather and analyze data about their communities. Suffrage maps were crafted, distributed, and displayed widely across the country, beginning in 1908 and continuing until women received the vote in 1920. The National American Woman Suffrage Association created a publishing company to keep up the national (and at times international) demand for suffrage propaganda, including the suffrage map which had to be updated and reprinted with each success. In parades and pageants, suffragists went a step further beyond displaying the map, becoming a "living suffrage map" itself, with robed women representing the states, their dress reflecting the states' suffrage status. Antilynching activists created and circulated widely maps depicting lynching by state to bring greater public attention and pressure on those states with higher numbers of lynchings, predominately in the South. Many of the maps created by American women during this era were maps documenting changes they sought to make in the landscape – be it new highways or the end of practices such as lynching. That is to say, they were in "motion" also. As the suffrage Literature Committee grumbled: "varium et mutabile semper *literature* est" – "literature is always varied and changeable." Like suffrage materials, the geographies and maps being created and circulated by American women in the Progressive Era were always in process, in motion.

As American women used and created geographic knowledge and maps, they were engaged in a practice of public geography. Through basic

American education and, at times, higher education, women *knew* geography and maps. They knew about the discipline of geography, about thinking spatially, about how to read maps, about how to use maps. As they became more mobile and active in the public sphere, they came to employ maps in a wide-variety of ways, from wayfinding to the persuasion of male voters. Not only did they create maps, they crafted their maps to heighten their message.

This practice of geography and cartography is a distinctive women's mapping tradition: they were creating geographies and crafting maps for their own purposes. With the bicycle and automobile, they addressed distinctive women's concerns about mobility, about how to dress, how it was possible to be more mobile and still be proper, and where it was appropriate to go. With the automobile, they went a step further to argue for the marking of historic trails/roads and the creation of new roads, going beyond women's needs and concerns to issues of importance to the nation such as the importance of a strong infrastructure or the importance of reducing rural isolation. With missionaries, they were educating their audiences on their mission communities and promoting geographic education in their efforts. While much of their work was fairly conventional facts about countries and cultures, there was a strong emphasis on topics and subjects important to women. As they discussed the countries and cultures, they included information about the status of women and children, providing primers on practices such as footbinding, female infanticide, zenana, and suttee. They drifted into propaganda in their efforts to promote missionary work in Africa (the "Dark Continent") or in attempting to mobilize the American public against the Mormon "threat." And maps were part of all these missionary efforts.

In another form of "women's work," social settlement workers sought to make a difference in their country while finding meaningful work for themselves. At their settlement houses, they provided English language classes, childcare, educational opportunities as well as space for intellectual and social engagement, and at times medical clinics, employment offices, and relief bureaus. As they worked to improve their communities, community studies were very much a part of the social settlement approach, studying their communities to better understand them and their needs. While useful to understand the community and its needs, these "social studies" were crucial to providing persuasive evidence of the issues as they approached local governments about addressing issues such as sanitation, the treatment of juveniles, or the legal rights of women workers and mothers, using the power of statistics and maps to provide persuasive evidence of the problems. Where it was appropriate, they mapped their data and used the maps to argue for changes. And American women lobbied for legislative change. With suffrage, it was to have a say in their country and have greater control over the laws that impacted them and their children, with the suffrage map becoming an important component of their rhetoric in arguing for the vote. With lynching, it was to address a national social issue that they felt deeply had to change and which they felt was being intentionally overlooked legally by the hegemonic white male political machine. In this case also, the map became part of the rhetoric employed to persuade their audiences of the need for change.

Was this a distinctive women's geographic and mapping tradition? I believe yes. Most of the geographic knowledge and maps they circulated were fairly straightforward and not unlike the male-generated geographic knowledge and maps widely circulating in American culture. But the focus on providing for their women-specific audience was striking, whether it was appropriate cycling locations, information about their efforts to end foot-binding, or how to make a large suffrage map for a rally. The maps created for the suffrage and anti-lynching campaigns were predominantly for a male voter audience, but these actions by women's organizations to influence male voting behavior were unprecedented in the United States, I believe. They were employing the masculinist discourse of geography and maps for their own ends, and towards persuading a male audience. While many of the maps they created were conventional, not visually remarkable, and didn't contribute to the canon of geography, the suffrage maps in particular represent the conscious molding of the map into a sophisticated tool of rhetoric that was used extensively, not only across the country but also over a twelve-year period, as they pressed the suffrage question from state to state until they achieved their goal. It does indeed appear that Progressive era women were well aware of the "power and the pleasure" of maps (Huffman 1997, 255).

When I began this research in 2002, I was interested in the question "when did women become part of the map reading audience?" How was it that women were gazing in pleasure at maps in the 1930s? I feel I now have the answer. Over the course of the nineteenth century, American women with access to education were exposed to geography and maps. By the turn-of-the-twentieth century, women had shifted from passive consumers of maps to practitioners of geography and cartography, creating and circulating their own geographies and maps. By the 1930s, American women were well acquainted with geography and maps. Not only could they now gaze with pleasure at maps and plot their next adventure (Figure 1.1), they could themselves make maps and use them in their activities (examples of maps throughout the book).

There is no definitive "end" to the Progressive Era. Although many scholars use 1920 as an approximate end, women's work did not end in 1920 with many of the organizations, associations, networks continuing (Perry 2002, 43–46). American women had considerable success during the Era on a wide variety of fronts, yet their success led to societal changes that, while better for American society, meant less "women's work" than they had been engaged in. Their efforts to improve sanitation, food quality, treatment of juveniles had led to the creation of agencies to address these issues, but now largely handled by men although some women found work there. Women's missionary work led to the merging of men and women's missionary forces into a larger, combined effort but more often led by men and leading overall

to a decline in missionary endeavors. Women received the vote and suffragists shifted their focus to educating and supporting voters, especially
women. Some suffragists shifted their efforts to the antilynching movement,
until that declined and the war effort surged around 1941.

But women's interest in and knowledge of geography and cartography did
find new outlets. During World War II, women were pressed into the war
effort on a wide variety of fronts. Besides entering manufacturing in traditional male roles (i.e. "Rosie the Riveter"), they took on traditional male
roles in geography and cartography. Given the need for spatial information
during the war and the lack of sufficient coverage, the US government created training programs at colleges specifically in drafting, surveying and
photogrammetry (Anderson 2014; Tyner 1999, 23). The expectation had
been that men would take these courses, instead significant numbers of
women enrolled and were put to work. Judith Tyner documents the work
of American women in U.S. government mapping offices such as the Office
of Strategic Services (OSS), the Tennessee Valley Authority, and the U.S.
Coast and Geodetic Survey in what she terms "Millie the Mapper" (Tyner
1999). "Military mapping maidens" were used by the Army Map Service for
a wide variety of mapping assignments during the war (Anderson 2014).[4]
In addition to mapping assignments, women geographers worked in a wide
variety of US government agencies. Chauncy Harris, in his account of his
work in Washington D.C. during World War II, names no less than fourteen
women who were among the ranks of geographers in Washington (Harris
1997). Unfortunately, when American men returned from war, the women
were expected to happily return to their homes and their lives as wives and
mothers. But not all women were content with these roles: some stayed on in
government positions (Andrews 1989; Harris 1997, 253).

The most challenging aspect of this project was the breadth and volume
of the material but also the difficulty at times in locating specific maps.[5]
When I began this research in 2002, I was just looking at road maps and
wondering about women using maps. From here, I began considering (at
the advice of James Akerman of The Newberry Library), women and bicycling as a possible start to women and map-use. While researching women
and bicycling at the American Geographical Society Library (University of
Wisconsin-Milwaukee), I saw their poster-sized suffrage maps and had to
learn more. While researching suffrage maps and trying to locate the first
suffrage map in Chicago, I learned more about the efforts of Hull-House,
their studies, and their maps. Social settlement efforts led me to consider
missionary efforts. I feel I have been following threads, with one map or
map reference leading to another.

The most frustrating aspect of this work are the unlocateable maps. The
maps I have been pursuing are not "precious" maps: they are mass produced
and widely distributed, cheaply made and quickly dated. They were not to
be cherished but to be used, hopefully lead to changes in society and then
tossed, disposable maps once their purpose has been served. As a result, I

have a wish list of maps that I am still trying to locate. I have just enough information to be certain of their existence but have not, to date, located an extant copy, including: Santa Fe Trail map created by the DAR of Kansas, WTUL studies from 1905 to 1906, 1st suffrage map and several later suffrage maps, "Mrs. Armour's maps" (created and sold in Illinois to raise funds for the World War I effort and used at home to track war movements), maps used by the President of the WCTU, and more. I still hope that these maps might be located, studied, and published, enriching our understanding of women's use of geography and mapping in the Progressive Era.

It is hard not to dwell on an alternative reason why these maps were not preserved: that women's geography and mapping was not deemed "worthy" of preservation. Their work, outside of the academy and widely accessible, did not make significant contributions to the canon of geographical knowledge generated by primarily by academic geographers. It rarely is acknowledged in histories of geography. Karen Morin, in her work on geographer Charles Daly reflects on how histories and historiographies of geography produce geography:

> ... histories and historiographies of geography *produce* the realm or limits of geography as much as they reflect some pre-existing condition of it as a coherent, mutually agreed upon body of knowledge, with particular protagonists central to the narratives. Who becomes a major and minor figure, who has the authority to speak, what kind of knowledge and skills make for dominant traditions within our discipline, and so on are important questions about the social construction of geographical knowledge and the 'worlding' of its creators.
>
> Morin 2011, 202[6]

Geographers and cartographers working in academia and government agencies would have their work largely preserved in libraries and eventually shifted into archives. With governmental work, it was kept as part of the "official" record of governmental actions. With academic work, it was often created as part of the academic enterprise of pursuing knowledge and truth. In both cases, this work was valued and preserved. This was not the intent of "women's work." They were not interested in creating geographies and maps to stand the test-of-time but were rather working to change their landscapes as soon as possible. As Avril Maddrell discusses in her book *Complex Locations*, in order to illuminate the contributions of women and their practices of geography, it is a process of "excavating and explicating," using archives, oral histories, obituaries, reviews of publications, and whatever else can be found, in order to bring out these "hidden histories" (Maddrell 2009, 6). Because the majority of these women were operating outside of academia and outside of government, their efforts are minimally documented. Some women had the opportunity to join institutions but chose not to. Jane Addams had the opportunity to join the University of Chicago, to become part of the academic world, but chose not to affiliated herself or Hull-House. The concerns of Addams and Hull-House were more immediate: improving as soon as possible the lives of Chicago's South Side residents. My unlocateable maps may be out there, tucked into a scrapbook or a collection of papers, but so far they have remained elusive.

But as I follow my threads and pursue maps, I have found other maps that similarly have had little serious consideration and offer new avenues of research. A chance glance at a news website over lunch introduced me to the Tuskegee Institute's map of lynchings [1931?].[7] Attempting to learn more about this map led to tripping over the Association of Southern Women for the Prevention of Lynching's maps as well as the NAACP maps. In trying to learn more about these anti-lynching maps, I learned of Nathan Work's publications out of Tuskegee and his use of maps in his publications such as the *Negro Year Book*.[8] I am beginning to collect evidence of what I would consider true subaltern counter-public mapping by Black Americans in the Progressive Era, yet another set of figures in this game of cat's cradle. I hope to explore this further in the near future.

What I have sketched out here is just one possibility in the game of cat's cradle I have been engaged in. In approaching women's use and production of geography and maps in the Progressive Era, I have been working with a variety of threads to create one possible pattern/interpretation of women's geographic work. Another scholar would likely produce at least a slightly different pattern/interpretation. Nevertheless, I believe it is time we recognize the practice of geography and mapping by American women in the Progressive Era. They may not have transformed geography but they did significantly contribute to the evolution of the American landscape. This pattern cannot be ignored as it is strong and the threads numerous. But then the question arises: what other practices/patterns have been ignored? I hope to explore one that I have begun to put into play, a thread found in the course of this work, but others are likely out there. And I look forward to seeing what patterns/practices future scholars bring forward, enriching our history of geography so that it truly become histo*ries* of geography.

Notes

1 *Good Housekeeping* July 1928, pp. 108.
2 Latin translation: "literature is always varied and changeable." Emphasis is from Livermore.
3 My conceptualization of "maps in motion" which underlines much of my work here originates with the late Denis Cosgrove's work in *Geography and Vision* (2008), specifically his chapter on "Moving Maps" where he discusses the shifting landscape of maps and cartography. Cosgrove has long inspired me academically and when I encountered his phrase "moving maps," it struck me for the first time how important the concept of mobility was to the work I was doing on Progressive Era American women's work in geography and cartography.
4 The term "Millie the Mapper" was coined by Tyner in her 1999 article. "Military mapping maidens" was devised and used by the Army Map Service during

World War II. At times, it was abbreviated to "3Ms" – "They called us the 3Ms, (and) we thought it was appropriate and rather cute" Bea McPherson quoted in Anderson (2014).

5 Geographers have long been called "jacks-of-all-trades and masters of none" – I certainly felt that deeply as I approached the various realms where women employed geography and maps in the Progressive Era!

6 Emphasis is Morin's.

7 In January 2013, the online magazine *Slate* featured the Tuskegee Institute's map of lynchings on their history blog "The Vault." See: www.slate.com/blogs/the_vault/2013/01/08/lynching_map_tuskegee_institute_s_data_on_lynching_from_1900_1931.html. Accessed 20 May 2017.

8 Work and his associates published the *Negro Year Book* sporadically, producing eleven editions between 1912 and 1952 (McMurry 1980, 339).

Bibliography

Anderson, A. 2014. WWII 'mapping maidens' chart course for today's mapmakers. *NGA Pathfinder* 12, 1: 14–17.

Andews, A. 1989. Women in applied geography. In *Applied Geography: Issues, Questions and Concerns*, ed. M. Kenzer, pp. 193-204. Dordrecht: Kluwer Academic Publishers.

Cosgrove, D. 2008. *Geography and Vision: Seeing, Imagining and Representing the World*. London: I.B. Tauris.

Harris, C. 1997. Geographers in the U.S. Government in Washington, D.C. in World War II. *Professional Geographer* 49, 2: 245–256.

Huffman, N. 1997. Charting the other maps: cartography and visual methods in feminist research. In *Thresholds in Feminist Geography: Difference, Methodology, Representation*, eds. J. Jones III, H. Nast, and S. Roberts, pp. 255–283. Lanham, MD: Rowman & Littlefield Publishers, Inc.

Livermore, H. 1919. Report of the literature committee. In *Handbook of the National American Woman Suffrage Association and Proceedings of the Jubilee Convention*, ed. J. Wilson, pp. 171–174. New York: National American Woman Suffrage Association.

Maddrell, A. 2009. *Complex Locations: Women's Geographical Work in the U.K. 1850–1970*. Chichester: Wiley-Blackwell.

McMurry, L. 1980. A black intellectual in the New South: Monroe Nathan Work, 1866–1945. *Phylon* 41, 4: 333–344.

Morin, K. 2011. *Civic Discipline: Geography in America, 1860–1890*. Farnham, UK: Ashgate Publishing Ltd.

Perry, E. 2002. Men are from the gilded age, women are from the progressive era. *Journal of the Gilded Age and Progressive Era* 1, 1: 25–48.

Schnell, I. and C. Leuenberger. 2014. Mapping genres and geopolitics: the case of Israel. *Transactions of the Institute of British Geographers* NS 39, 4: 518–531.

Tyner, J. 1999. Millie the mapper and beyond: the role of women in cartography since World War II. *Meridian* 15: 23–28.

Index

"see one, seen them all" attitude 181
semiotic homogeneity 110
Semple, Ellen Churchill 155, 157
Seneca Falls Women's Rights
   Convention 176
"separate spheres" 33, 84, 194, 225
settlement work: activities 120; cultural
   capital 124; implications, geography/
   mapping 159–62; mapping, affiliated
   organizations 145–53; mapping as
   social survey work 161; overview 23–4,
   81, 119–20, 226; parallels to missionary
   work 160; women's work 159, 160–2;
   see also Hull-House settlement
shackles 185, 195, 200, 202
Shaw, Anna Howard 203
Siam missionaries 88
Simkhovitch, Mary 141
situated knowledge 24, 145, 159
slavery 192–5
Smith, Catherine Delano 25
Smith, Joseph 104
Smith College 4
soapbox speakers 198, 202
social activism 24, 138, 145–53, 173;
   see also specific area of activism
social centers 19
social conditions, mapping 123
social geography 11, 120, 156–7
"social gospel" movement 83–4, 120
social problems 83
social reforms movements 176
social research 141
"The Social Research Map" 141
social settlement 23, 84, 119–21, 133,
   139, 140, 144, 157, 160, 161, 162, 175,
   193, 226, 227, 229
social surveys 120, 141–2
sociological surveys 141
South America missionaries 88
Southern Baptist Missionary Union 210
Southern Commission on the Study of
   Lynching 206
Southern Good Roads 63
Southern mission work 88
affinity spaces 19; automobile and 48,
   69, 71; concept of 8, 11, 34, 48, 71,
   156, 187; domestic space 88, 176, 194;
   feminism and 17, 20; gendered space
   178, 225; map space 6, 180; mission
   space 89, 92, 107; public space 35, 40,
   72, 178,194, 197, 200, 215; race and
   47; subaltern and 18–9; women and
   control of 24, 35, 69, 81

Speier, Hans 175, 204
Springer, Helen Emily 88
social geography 11, 120, 156–7
staged tableaus 200
Standard Oil Company 1
Standard Runabout 53
Stanley, Henry Morton 103, 193
Stanton, Elizabeth Cady 46, 49, 177
Starley, John Kemp 35
Starr, Ellen Gates 5, 82, 119, 121–2, 133
State Federation of Women's
   Clubs 210
Stead, William 121, 193
Strabo (Greek geographer) 46
"subaltern cartographies" 18, 25
subaltern counter-public mapping 231
subaltern counter-publics 18
Su Dejeng, Wang 95–6
suffrage movement: after passage 203;
   anti-suffrage rhetoric 194; assault on
   women 199, 202; autotours 56–7; black
   men allowed to vote 177; cloudless
   map 203–5; color use 179, 185;
   Congressional hearings 199; dark spot
   on the map 191–5; elements applied
   177–8; gendered differences 173, 177,
   185–6; "Hopperie Game" 201–2; living
   map 195–202; maps and 176–205;
   moral maps 182–5; overview 173–6,
   226, 227; police assault on women
   199; public presence 177–8, 195–202;
   regional suffrage maps 189; reinforcing
   women's place in society 178; resisting
   the map 202–3; school suffrage 177,
   180, 185, 203; soapbox speakers 198,
   202; suffrage map conflict 179–82;
   visibility tactics 177–9
summer mission schools 92
Swain, Clara 89
sweating system 131–2

Teape, Nancy 55, 60
Teape, Vera Marie 55, 60
telephone/telegraph poles, painting 66
'Temple of Time' 4
Temporary Autonomous Zones (TAZs) 19
Terrell, Mary Church 207
Theatrum Orbis Terrarum 200
thematic maps/mapping 7–10, 105, 107,
   123, 159, 174, 175, 183, 205, 215
*Thirty Years of Lynching in the United
   States 1889–1918* 191
Through the Dark Continent 103, 193
Tinnin, Glenna Smith 195–6, 200

Milton Keynes UK
Ingram Content Group UK Ltd.
UKHW040107071024
449327UK00019B/879

9 780367 245306